Winning Ways
for Your Mathematical Plays

Volume 3, Second Edition

Elwyn R. Berlekamp, John H. Conway, Richard K. Guy

A K Peters
Natick, Massachusetts

Editorial, Sales, and Customer Service Office

A K Peters, Ltd.
63 South Avenue
Natick, MA 01760

Library of Congress Cataloging-in-Publication Data

Berlekamp, Elwyn R.
   Winning Ways for your mathematical plays / Elwyn Berlekamp, John H. Conway,
   Richard Guy.--2nd ed.
      p. cm.
   Includes bibliographical references and index.
   ISBN 1-56881-130-6 (v. 1) – ISBN 1-56881-142-X (v. 2) – ISBN 1-56881-143-8 (v. 3) –
   ISBN 1-56881-144-6 (v. 4) (alk.paper)
   1. Mathematical recreations. I. Conway, John Horton. II. Guy Richard K. III. Title.

QA95 .B446 2000
739.7'4--dc21   00-048541

Printed in Canada

07 06 05 04 03   10 9 8 7 6 5 4 3 2 1

To Martin Gardner

who has brought more mathematics to more millions than anyone else

Elwyn Berlekamp was born in Dover, Ohio, on September 6, 1940. He has been Professor of Mathematics and of Electrical Engineering/Computer Science at UC Berkeley since 1971. He has also been active in several technology business ventures. In addition to writing many journal articles and several books, Berlekamp also has 12 patented inventions, mostly dealing with algorithms for synchronization and error correction.

He is a member of the National Academy of Sciences, the National Academy of Engineering, and the American Academy of Arts and Sciences. From 1994 to 1998, he was chairman of the board of trustees of the Mathematical Sciences Research Institute (MSRI).

John H. Conway was born in Liverpool, England, on December 26, 1937. He is one of the preeminent theorists in the study of finite groups and the mathematical study of knots, and has written over 10 books and more than 140 journal articles.

Before joining Princeton University in 1986 as the John von Neumann Distinguished Professor of Mathematics, Conway served as professor of mathematics at Cambridge University, and remains an honorary fellow of Caius College. The recipient of many prizes in research and exposition, Conway is also widely known as the inventor of the Game of Life, a computer simulation of simple cellular "life," governed by remarkably simple rules.

Richard Guy was born in Nuneaton, England, on September 30, 1916. He has taught mathematics at many levels and in many places— England, Singapore, India, and Canada. Since 1965 he has been Professor of Mathematics at the University of Calgary, and is now Faculty Professor and Emeritus Professor. The university awarded him an Honorary Degree in 1991. He was Noyce Professor at Grinnell College in 2000.

He continues to climb mountains with his wife, Louise, and they have been patrons of the Association of Canadian Mountain Guides' Ball and recipients of the A. O. Wheeler award for Service to the Alpine Club of Canada.

Contents

Preface to the Second Edition

In the first edition of *Winning Ways*, which appeared in 1982, we were able to make a rather sharp distinction between those games in Part I, to which the major theory of addition applied directly, and those games in Part 3, which seemed to require more specialized techniques. However, subsequent research by an increasingly large community of combinatorial game theorists has begun to blur this distinction. We now have many more games whose strategies depend both on the general theory of Volume 1 as well as on more specialized results. Introductions to many of these games and some illustrative problems have been added to this new edition. Those that did not readily fit elsewhere can be found in the new Extras to Chapter 22 at the end of this volume. This volume also includes a major revision of the original Chapter 20 on the game of Fox and Geese. Its enhanced variation, Fox-Flocks-Fox, provides compelling illustrations of some of the challenging problems that can now be solved by appropriately combining theories from Volumes 1, 2, and 3 with innovative computing algorithms.

This new edition owes much to the supportive efforts of numerous friends and colleagues, including Noam Elkies, Tom Ferguson, Aviezri Fraenkel, Martin Gardner, Sol Golomb, Al Hales, Greg Kuperberg, Silvio Levy, Donald Knuth, Martin Kutz, Greg Martin, Victor Meally, Richard Nowakowski, Hilarie Orman, Marc Paulhus, Ed Pegg, Michael Reid, Thea van Roode, Katherine Scott, George Sicherman, Aaron Siegel, Neil Sloane, Sally Smith, William Spight, John Tromp, Jonathan Welton, Julian West, David Wilson, and David Wolfe, and to the very professional yet kindly support of our publishers, Alice and Klaus Peters.

Elwyn Berlekamp, University of California, Berkeley
John Conway, Princeton University
Richard Guy, The University of Calgary, Canada

June 23, 2003

Preface to the Original Edition

Does a book need a Preface? What more, after fifteen years of toil, do three talented authors have to add.

We can reassure the bookstore browser, "Yes, this is just the book you want!"

We can direct you, if you want to know quickly what's in the book, to page xx. This in turn directs you to volumes 1,2,3 and 4.

We can supply the reviewer, faced with the task of ploughing through nearly a thousand information-packed pages, with some pithy criticisms by indicating the horns of the polylemma the book finds itself on. It is not an encyclopedia. It is encyclopedic, but there are still too many games missing for it to claim to be complete. It is not a book on recreational mathematics because there's too much serious mathematics in it. On the other hand, for us, as for our predecessors Rouse Ball, Dudeney, Martin Gardner, Kraitchik, Sam Loyd, Lucas, Tom O'Beirne and Fred. Schuh, mathematics itself is a recreation. It is not an undergraduate text, since the exercises are not set out in an orderly fashion, with the easy ones at the beginning. They are there though, and with the hundred and sixty-three mistakes we've left in, provide plenty of opportunity for reader participation. So don't just stand back and admire it, work of art though it is. It is not a graduate text, since it's too expensive and contains far more than any graduate student can be expected to learn. But it does carry you to the frontiers of research in combinatorial game theory and the many unsolved problems will stimulate further discoveries.

We thank Patrick Browne for our title. This exercised us for quite a time. One morning, while walking to the university, John and Richard came up with "Whose game?" but realized they couldn't spell it (there are three tooze in English) so it became a one-line joke on line one of the text. There isn't room to explain all the jokes, not even the fifty-nine private ones (each of our birthdays appears more than once in the book).

Omar started as a joke, but soon materialized as Kimberly King. Louise Guy also helped with proof-reading, but her greater contribution was the hospitality which enabled the three of us to work together on several occasions. Louise also did technical typing after many drafts had been made by Karen McDermid and Betty Teare.

Our thanks for many contributions to content may be measured by the number of names in the index. To do real justice would take too much space. Here's an abridged list of helpers: Richard Austin , Clive Bach, John Beasley, Aviezri Fraenkel, David Fremlin, Solomon Golomb, Steve Grantham, Mike Guy, Dean Hickerson, Hendrick Lenstra, Richard Nowakowski, Anne Scott, David Seal, John Selfridge, Cedric Smith and Steve Tschantz.

No small part of the reason for the assured success of the book is owed to the well-informed and sympathetic guidance of Len Cegielka and the willingness of the staff of Academic Press and of Page Bros. to adapt to the idiosyncrasies of the authors, who grasped every opportunity to modify grammar, strain semantics, pervert punctuation, alter orthography, tamper with traditional typography and commit outrageous puns and inside jokes.

Thanks also the the Isaak Walton Killam Foundation for Richard's Resident Fellowship at The University of Calgary during the compilation of a critical draft, and to the National (Science & Engineering) Research Council of Canada for a grant which enabled Elwyn and John to visit him more frequently than our widely scattered habitats would normally allow.

And thank you, Simon!

University of California, Berkeley, CA 94720 Elwyn Berlekamp
University of Cambridge, England, CB2 1SB John Conway
University of Calgary, Canada, T2N 1N4 Richard Guy

You are
now here

If you want to know roughly what's elsewhere,
turn to the little notes about our four main themes:

There are a number of other connections between various chapters of the book:

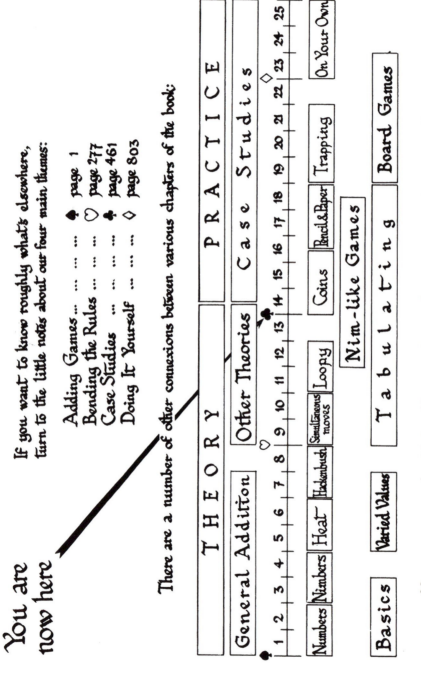

However, you should be able to pick any chapter and read almost all of it
without reference to anything earlier, except perhaps the basic ideas at the start of the book.

Games in Clubs!

To be an Englishman is to belong
to the most exclusive club there is.
Ogden Nash, *England Expects.*

There are lots of games for which the theories we've now developed are useful, and even more for which they're not, and we've grouped them into clubs according to how you play them.

First some games you can play with coins, either by turning them over (Chapter 14) or moving them along strips or about in heaps (Chapter 15).

Then games for which you'll need pencil and paper, perhaps to draw straight lines (Chapter 16), or curved ones (Chapter 17) or merely to do the calculations in Chapter 18.

And for board games we have three case studies in which one player wins by trapping his opponent (Chapters 19, 20, 21) and finally many more which are usually won by the first player to establish some kind of winning configuration (Chapter 22).

-14-

Turn and Turn About

Because I do not hope to turn again
Because I do not hope
Because I do not hope to turn.
        T. S. Eliot, *Ash Wednesday*, I.

Open not thine heart to every man, lest he requite thee
with a shrewd turn.
        Ecclesiasticus, 8:19.

These games, based on an idea of H. W. Lenstra, are similar in that they all involve turning things over, but we shall see that they call for a variety of strategies.

Turning Turtles

Figure 1. Playing Turning Turtles.

461

In Fig. 1 the Walrus and the Carpenter are playing a rather cruel game. At each move a player must put one turtle on its back and may also turn over any single turtle to the left of it. This second turtle, unlike the first, may be turned either onto its feet or onto its back. The player wins who turns the last turtle upside-down. Which turtles should the Walrus ($l.$) turn?

Like most readers of this book, he wearily suspects another disguise for Nim. Here only turtles 3, 4, 6, 8 and 10 are on their feet, and since the nim-sum of 3, 4 and 6 is 1, he may turn 10 onto its back and 9 onto its feet, producing 3, 4, 6, 8, 9, a $\mathcal{P}$-position since $8 \overset{*}{+} 9 = 1$. The Carpenter ($r.$)responds by turning 8 and 5 producing the position 3, 4, 5, 6, 9 as in Fig. 2.

Figure 2. After the Carpenter's Reply.

In Nim there is only one good move from this position—reduce 9 to 4, so as to produce, 3, 4, 4, 5, 6, which, since two equal Nim heaps may be cancelled, is much the same as 3, 5, 6, which the Walrus reaches by turning both 9 and 4 on their backs (Fig. 3).

Figure 3. How the Walrus Won.

Nim moves become turtle turns as follows. We reduce a heap to a size not already present by turning one turtle on its back and putting another on its feet, as in the Walrus's opening move. If a heap of the reduced size is already present, we turn two turtles on their backs as in the Walrus's response to the Carpenter's move (cancelling two equal heaps). To eliminate a heap entirely, we merely turn the appropriate turtle. So since 4, 6, 8, 10 is a $\mathcal{P}$-position, the Walrus could have won from Fig. 1 by just turning turtle 3.

Since all our turning games are impartial, they are solved by computing the nim-values, and often may be thought of as heap games in disguise; but many games with interesting theories are more naturally suggested by the turning version.

Figure 4. The Mock Turtle Joins in.

Mock Turtles

Let the players turn up to three turtles subject only to the condition that the rightmost of these must be turned from his feet onto his back. We may think of this as a game with numbers in which any number may be replaced by 0, 1 or 2 smaller ones. So $\mathcal{G}(n)$ is the least number not of any form

$$0, \mathcal{G}(a), \mathcal{G}(a) \overset{*}{+} \mathcal{G}(b),$$

in which a and b are any numbers less than n.

If we number the positions from 0, we find the nim-values shown in Table 1.

$$n = 0\ 1\ 2\ 3\ 4\quad 5\quad 6\quad 7\quad 8\quad 9\ 10\ 11\ 12\ 13\ 14\ 15\ 16\ 17\ 18 \ldots$$
$$\mathcal{G}(n) = 1\ 2\ 4\ 7\ 8\ 11\ 13\ 14\ 16\ 19\ 21\ 22\ 25\ 26\ 28\ 31\ 32\ 35\ 37 \ldots$$

Table 1. Nim-values for Mock Turtles.

We see that $\mathcal{G}(n)$ is always $2n$ or $2n+1$, so that its binary expansion is obtained by adjoining a digit 0 or 1 to that of n. Which shall it be?

$$n = 0\quad 1\quad 10\quad 11\quad 100\quad 101\quad 110\quad 111\quad 1000\quad 1001\quad 1010 \ldots$$
$$\mathcal{G}(n) = \underline{1}\ 1\underline{0}\ 10\underline{0}\ 11\underline{1}\ 100\underline{0}\ 101\underline{1}\ 110\underline{1}\ 111\underline{0}\ 1000\underline{0}\ 1001\underline{1}\ 1010\underline{1} \ldots$$

Table 2. The Odious Numbers Revealed.

Table 2 suggests we choose whichever makes the total number of 1-digits *odd*.

Odious and Evil Numbers

Every number is **odious** or **evil** according to the number of 1's in its binary expansion (odious for odd, evil for even). These behave under Nim addition like odd and even numbers under ordinary addition:

$$\text{EVIL} \overset{*}{+} \text{EVIL} = \text{EVIL} = \text{ODIOUS} \overset{*}{+} \text{ODIOUS},$$
$$\text{EVIL} \overset{*}{+} \text{ODIOUS} = \text{ODIOUS} = \text{ODIOUS} \overset{*}{+} \text{EVIL}.$$

When we compute $\mathcal{G}(n)$ in Mock Turtles, the next odious number is *never* excluded, because the nim-sum of two odious numbers is evil, but smaller evil numbers always *are* excluded.

If $a_1, a_2, \ldots, a_n$ is a $\mathcal{P}$-position in Nim, so that

$$a_1 \overset{*}{+} a_2 \overset{*}{+} \ldots \overset{*}{+} a_n = 0,$$

then for the corresponding odious numbers $\mathcal{G}(a_i)$ in Mock Turtles we shall have

$$G(a_1) \overset{*}{+} G(a_2) \overset{*}{+} \ldots \overset{*}{+} G(a_n) = 0 \text{ or } 1.$$

But if n is even, this nim-sum is evil, and so 0; while if n is odd it is odious, and so 1. The $\mathcal{P}$-positions in Mock Turtles are therefore just those $\mathcal{P}$-positions in Nim for which n is even.

Note that in Mock Turtles we number the turtles from 0. The turtle numbered 0, called the Mock Turtle, must take his turn with the rest and cannot be neglected in the conversion to Nim. To obtain a $\mathcal{P}$-position in Mock Turtles from the Turning Turtles position of Fig. 3, the Mock Turtle must be brought into the game with his four feet on the ground. In Mock Turtles, 3, 5, 6, is *not* a $\mathcal{P}$-position, but 0, 3, 5, 6 is (Fig. 4).

Moebius, Mogul and Gold Moidores

Table 3 shows the nim-values, kindly checked for us on the computer by M.J.T. Guy, for similar games in which we may turn over up to t objects for $t = 1, 2, \ldots$. Because the numbers get much larger than the other nim-values in this book, we have written them in base 8 (octal) notation. Nim-sums of octal numbers may be computed digit by digit thus:

$$
\begin{array}{r}
1\,2\,3\,4\,5\,6\,7\,0 \\
1\,3\,5\,7\,0\,2\,4\,6 \\
\hline
1\,6\,3\,5\,4\,3\,6.
\end{array}
$$

In the table we have only named the most interesting cases: $t = 3, 5, 7$ and 9. Note that C, E, G and I are the 3rd, 5th, 7th and 9th letters of the alphabet. For convenience, and to avoid cruelty to turtles, the reader may play these games with coins. The coins will show heads or tails according as the turtle is on his feet or on his back, and the rightmost coin that is turned must change from heads to tails.

The Mock Turtle Theorem

Take a $\mathcal{P}$-position in the game for an even value of $t, t = 2m$, and place an extra coin (the Mock Turtle) at the left, whichever way up will ensure an even number of heads. Positions obtained in this way will be called "good" positions for the next odd value of t, $t = 2m + 1$. We assert that the good positions are precisely the $\mathcal{P}$-positions for the game $t = 2m + 1$.

We show first that there is no way of changing from one good position to another by turning at most $2m + 1$ coins. If there were, the number of coins turned would necessarily be even, since the good positions have evenly many heads, and so would actually be at most $2m$. But this would entail a move between two $\mathcal{P}$-positions in the $2m$ game.

n	t = 1	2	3	4	5	6	7	8	9
		MOCK TURTLES			MOEBIUS		MOGUL		MOIDORES
THE MOCK TURTLE	1	1			1		1		1
1	1	1	2	1	2	1	2	1	2
2	1	2	4	2	4	2	4	2	4
3	1	3	7	4	10	4	10	4	10
4	1	4	10	10	20	10	20	10	20
5	1	5	13	17	37	20	40	20	40
6	1	6	15	20	40	40	100	40	100
7	1	7	16	40	100	77	177	100	200
8	1	10	20	63	147	100	200	200	400
9	1	11	23	100	200	200	400	377	777
10	1	12	25	125	253	400	1000	400	1000
11	1	13	26	152	325	707	1617	1000	2000
12	1	14	31	200	400	1000	2000	2000	4000
13	1	15	32	226	455	1331	2663	4000	10000
14	1	16	34	253	526	1552	3325	7417	17037
15	1	17	37	333	667	1664	3551	10000	20000
16	1	20	40	355	733	2000	4000	20000	40000
17	1	21	43	367	756	2353	4726	31463	63147
18	1	22	45	400	1000	2561	5343	40000	100000
19	1	23	46	427	1056	2635	5472	52525	125253
20	1	24	51	451	1123	3174	6370	65252	152525
21	1	25	52	707	1617	3216	6435	100000	200000
22	1	26	54	1000	2000	3447	7116	113152	226325
23	1	27	57	1031	2063	3722	7644	200000	400000
24	1	30	61	1055	2132	4000	10000	213630	427461
25	1	31	62	1122	2245	10000	20000	263723	547646
26	1	32	64	1203	2407	20000	40000	306136	614274
27	1	33	67	1443	3106	34007	70017	400000	1000000
28	1	34	70	1537	3277	40000	100000	416246	1034515
29	1	35	73	1746	3714	54031	130063	521055	1242133
30	1	36	75	2000	4000	64052	150125	724616	1651435
31	1	37	76	2033	4066	70064	160151	1000000	2000000
32	1	40	100	2056	4134	100000	200000	1023305	2046613
33	1	41	103	2130	4261	114053	230126	1347214	2716431
34	1	42	105	2221	4443	124061	250143	2000000	4000000
35	1	43	106	2465	5153	130035	260072	2027151	4056322
36	1	44	111	2501	5203	144074	310170	2457261	5136542
37	1	45	112	3124	6250	150016	320035	3166444	6355111
38	1	46	114	3512	7225	160047	340116	4000000	10000000
39	1	47	117	4000	10000	174022	370044	4055666	10133554
40	1	50	121	4034	10071	200000	400000	4632577	11465377
41	1	51	122	4045	10113	214301	430603	5251417	12523036
42	1	52	124	4211	10423	224502	451205	7514712	17231625
43	1	53	127	4504	11211	230604	461411	10000000	20000000

Table 3. These Nim-values Are in Octal (base 8), *not* Decimal.

It remains to show that from any bad position in the $2m+1$ game there is a move to some good position. If the position is bad because it corresponds to an $\mathcal{N}$-position in the $2m$ game, there is a move in that game to some $\mathcal{P}$-position, and, by turning the Mock Turtle if necessary, we obtain a move to a good position in the $2m+1$ game. The other bad positions correspond to $\mathcal{P}$-positions in the $2m$ game, but have an odd number of heads. In this case, by turning over the rightmost head, we obtain a position that gives an $\mathcal{N}$-position in the $2m$ game. We can now turn over at most $2m$ further coins to make this a $\mathcal{P}$-position and then, if necessary to obtain a good position, also turn the Mock Turtle. We have turned at most $2m+2$ coins in all, but since we started with an odd number of heads and finished with an even number, we have in fact turned over at most $2m+1$ coins, and so have made a legal move in the $2m+1$ game.

This result is equivalent to the statement:

> Every nim-value for the $2m+1$ game is an odious number, and the corresponding value for the $2m$ game is obtained by dropping the final binary digit.

THE MOCK TURTLE THEOREM

Why Moebius?

$$\infty \quad 1 \quad 4 \quad 0 \quad -4 \quad -1 \quad 5 \quad 6 \quad -8 \quad 2 \quad -3 \quad -5 \quad 8 \quad 3 \quad -7 \quad 7 \quad -6 \quad -2$$

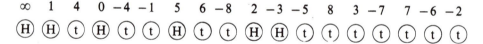

Figure 5. Moebius Labels Make $\mathcal{P}$-positions Easy to Find.

When restricted to 18 coins, the $\mathcal{P}$-positions of the game with $t = 5$ possess a remarkable symmetry. To see this, name the heads of a position by the numbers shown in Fig. 5. For example, the $\mathcal{P}$-position with heads in just the first 6 places is $\infty, 0, \pm 1, \pm 4$. In this notation $\mathcal{P}$-positions remain $\mathcal{P}$-positions when their numbers are increased by any fixed amount, modulo 17, leaving ∞ unchanged. Adding 1 to the numbers $\infty, 0, \pm 1, \pm 4$ we find $\infty, 1, 2, 0, 5, -3$, so that the position displayed in Fig. 5 is another $\mathcal{P}$-position. The 15 positions shown in Table 4 yield a total of $15 \times 17 = 255$ $\mathcal{P}$-positions in this way. It is also true that a $\mathcal{P}$-position remains a $\mathcal{P}$-position if we interchange heads and tails in every place. The positions with all tails or all heads are therefore both $\mathcal{P}$-positions, giving $2 \times 255 + 2 = 512$ $\mathcal{P}$-positions in all, distributed as follows:

Number of heads	0	6	8	10	12	18
Number of $\mathcal{P}$-positions	1	102	153	153	102	1.

6 heads			*8 heads*			
$\infty,0$	± 1	± 4	$\infty,0$	± 1	± 5	± 7
$\infty,0$	± 2	± 8	$\infty,0$	± 2	± 3	± 7
± 1	± 3	± 6	$\infty,0$	± 3	± 4	± 6
± 2	± 5	± 6	$\infty,0$	± 5	± 6	± 8
± 4	± 5	± 7	± 1	± 2	± 4	± 8
± 3	± 7	± 8	± 1	± 2	± 3	± 5
			± 2	± 4	± 6	± 7
			± 3	± 4	± 5	± 8
			± 1	± 6	± 7	± 8

Table 4. The $\mathcal{P}$-positions for Moebius.

Dropping the Mock Turtle (at ∞) we find that the $\mathcal{P}$-positions for the game $t = 4$ on 17 coins are distributed:

Number of heads	0	5	6	7	8	9	10	11	12	17
Number of $\mathcal{P}$-positions	1	34	68	68	85	85	68	68	34	1.

We can also double the numbers (modulo 17) of any $\mathcal{P}$-position to give another. Thus $\infty,0,1,2,-3,5$ of Fig. 5 becomes $\infty,0,2,4,-6,-7$. We can invert them modulo 17; since $1/2 = -8$, $1/3 = 6$ and $1/5 = 7$, Fig. 5 inverts into $0,\infty,1,-8,-6,7$. In fact we can make any transformation (modulo 17)

$$x \rightarrow \frac{ax+b}{cx+d}, \quad ad - bc = 1.$$

Since these are known as the Möbius transformations, we have named our game after that distinguished mathematician.

Mogul

On 24 coins the game for $t = 7$ displays even more symmetries. The $\mathcal{P}$-positions among the first 24 places are distributed as follows:

Number of heads	0	8	12	16	24
Number of $\mathcal{P}$-positions	1	759	2576	759	1.

Figure 6 enables us to find the 759 $\mathcal{P}$-positions with just 8 heads, or equally those with 8 tails. In either case the set of 8 places involved is called an **octad**. In Fig. 6 there are 35 **pictures** and each picture shows the 24 places colored in six sets of four (the 6 colors used are black, white, star, circle, plus and dot). Any two sets of 4 (any two colors) in the same picture make an octad: in particular this gives every octad with just 4 places in the last pair of (black and white) rows, and this pair of rows themselves form an octad. By interchanging this last pair of rows with the first pair, or the middle pair, of the same picture, we can now find all the octads, since it can be shown that these pairs of rows form octads and that every other octad meets at least one of them in just 4 places.

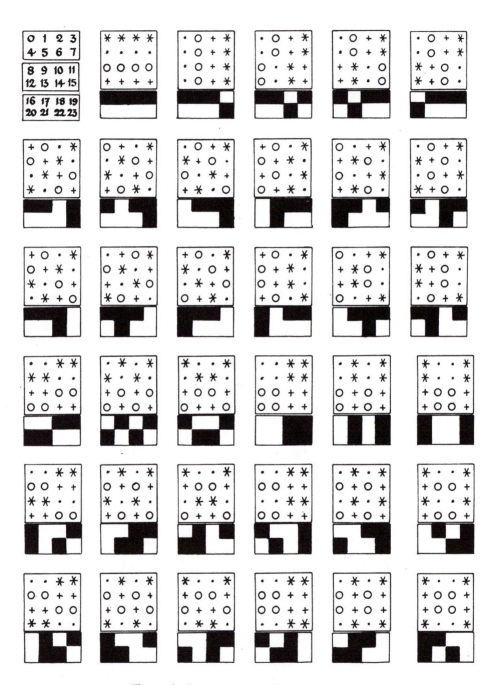

Figure 6. Curtis's Miracle Octad Generator.

This Miracle Octad Generator, or MOG, is due to R. T. Curtis, but we have modified it slightly for the Mogul player's convenience. Various regular features of its arrangement make it easy for the practised user to locate the unique octad containing any five given places. It seems to be the case that the winner in 24-place Mogul need never play into a 12-head $\mathcal{P}$-position.

Motley

This is the game in which any number of coins may be turned. When well played it lasts at most one move, since we can turn all the heads to tails instantly! The nim-values are the powers of 2:

$$1, 2, 4, 8, 16, 32, 64, 128, 256, 512, \ldots$$

so, when played with several rows, Motley is yet another disguise for Nim; the heads in a row are binary digits 1 in the number of beans in the corresponding Nim-heap.

Twins, Triplets, Etc.

We can also play the game **Twins**, in which we must turn *exactly* two coins, or **Triplets**, in which we turn exactly three, etc. The nim-value sequence for the game in which we turn exactly t coins consists of $t-1$ zeros followed by the nim-value sequence for the game in which we turn *at most* t coins. Thus the nim-values for Triplets are

$$0, 0, 1, 2, 4, 7, 8, 11, 13, 14, 16, 19, 21, 22, 25, \ldots.$$

We may think of the first $t-1$ coins as $t-1$ Mock Turtles which may be used to fill out our move to its proper complement of turns.

The Ruler Game

If the coins we turn must be *consecutive* but are otherwise unrestricted (except that the rightmost coin must be turned from heads to tails), then the nim-values are computed by the rule:

$$\mathcal{G}(n) = \text{mex} \left\{ \begin{array}{l} 0 \\ \mathcal{G}(n-1) \\ \mathcal{G}(n-1) \overset{*}{+} \mathcal{G}(n-2) \\ \mathcal{G}(n-1) \overset{*}{+} \mathcal{G}(n-2) \overset{*}{+} \mathcal{G}(n-3) \\ \cdots \cdots \end{array} \right\},$$

and are found to be reminiscent of Dividing Rulers (Fig. 7 of Chapter 13, Vol. 2).

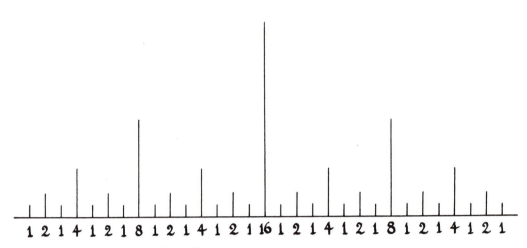

1 2 1 4 1 2 1 8 1 2 1 4 1 2 1 16 1 2 1 4 1 2 1 8 1 2 1 4 1 2 1

Figure 7. Nim-values for the Ruler Game.

If the coins are numbered starting from 1, $\mathcal{G}(n)$ is just the highest power of 2 dividing n.

Circumscribed Games

We can play any of these games under the additional restriction that the coins to be turned may not be too far apart. Thus in **Mock Turtle Fives** we may turn *up to three* of five consecutive coins. In **Triplet Fives** we turn *exactly three* out of five consecutive coins. In **Ruler Fives** we may turn 1, 2, 3, 4 or 5 consecutive coins. The nim-values for these three games are:

Mock Turtle Fives: 1 2 4 7 8 1 2 4 7 8 1 2 4 7 8 1 2 4 7 8 ...
Triplet Fives: 0 0 1 2 4 0 0 1 2 4 0 0 1 2 4 0 0 1 2 4 ...
Ruler Fives: 1 2 1 4 1 2 1 4 1 2 1 4 1 2 1 4 1 2 1 4 ...

These are parts of general patterns. Thus, **Moebius Nineteens**, for example, would have the first 19 values of Moebius repeated indefinitely. This happens for all the above games except the Ruler game; Ruler Fours, Sixes and Sevens have the same values as Ruler Fives, while Ruler Eights to Fifteens all have nim-values:

1 2 1 4 1 2 1 8 1 2 1 4 1 2 1 8 1 2 1 4 1 2 1 8 1 2 1 4 1 2 1 8 1 ....

Turnips (or Ternups)

This game has a richer theory, but it is a great pity that the full theory is only needed by people wealthy enough to play with a very large number of coins. The move is to turn over any three equally spaced coins, the rightmost going from heads to tails as usual. Numbering from 0 we find that the nim-values for 0 to 100 are:

0-8	0	0	1	0	0	1	2	2	1		
9-17	0	0	1	0	0	1	2	2	1		
18-26	4	4	1	4	4	1	2	2	1		
27-35	0	0	1	0	0	1	2	2	1		
36-44	0	0	1	0	0	1	2	2	1		
45-53	4	4	1	4	4	1	2	2	1		
54-62	7	7	1	7	7	1	2	2	1		
63-71	7	7	1	7	7	1	2	2	1		
72-80	4	4	1	4	4	1	2	2	1		
81-89	0	0	1	0	0	1	2	2	1		
90-100	0	0	1	0	0	1	2	2	1	4	4	...

Table 5. The Nim-values for Turnips.

To find $\mathcal{G}(n)$ in general, we expand n in base 3:

					n in ternary					$\mathcal{G}(n)$	
$\phi = 0$ or 1	...	ϕ	ϕ	ϕ	ϕ	ϕ	ϕ	ϕ		0	
$? = 0, 1$ or 2	...	$?$	$?$	$?$	$?$	$?$	$?$	2		1	the
	...	$?$	$?$	$?$	$?$	$?$	2	ϕ		2	odious
	...	$?$	$?$	$?$	$?$	2	ϕ	ϕ		4	numbers
	...	$?$	$?$	$?$	2	ϕ	ϕ	ϕ		7	in
	...	$?$	$?$	2	ϕ	ϕ	ϕ	ϕ		8	order
	...	$?$	2	ϕ	ϕ	ϕ	ϕ	ϕ		11	
	...	...	...	...	...	...	...	...		...	...

In words, $\mathcal{G}(n) = 0$ if the ternary expansion of n has no 2-digit, but is the kth odious number if the last 2-digit is in the kth place from the right, when we call n a k-**number**. The numbers n whose ternary expansions have no 2-digit will be called **empty numbers**.

To see all this, note that $\mathcal{G}(n)$ is the mex of all the numbers

$$\mathcal{G}(n - \delta) \overset{*}{+} \mathcal{G}(n - 2\delta) \text{ for } \delta = 1, 2, \ldots .$$

We show first that the putative value for $\mathcal{G}(n)$ is not one of these numbers, or equivalently that

$$\mathcal{G}(n) \overset{*}{+} \mathcal{G}(n - \delta) \overset{*}{+} \mathcal{G}(n - 2\delta) \neq 0.$$

Since the nim-sum of three odious numbers is odious, this will be true unless one of

$$\mathcal{G}(n), \mathcal{G}(n - \delta), \mathcal{G}(n - 2\delta)$$

is zero and the other two coincide. But if the last non-zero ternary digit ($x = 1$ or 2) of δ is in the kth place, the expansions of n, $n - \delta$, $n - 2\delta$ look like:

```
                k       j                                      k
δ:          ? ? ? x 0 0 0 0 0           δ:              ? ? ? x 0 0 0 0 0
──────────────────────────────         ──────────────────────────────────
   n  ⎞   ⎧ ? ? ? 0 ? ? ? 2 φ φ            n  ⎞   ⎧ ? ? ? 0 φ φ φ φ φ
 n−δ  ⎬:  ⎨ ? ? ? 1 ? ? ?,2 φ φ    or    n−δ  ⎬:· ⎨ ? ? ? 1 φ φ φ φ φ
 n−2δ ⎠   ⎩ ? ? ? 2 ? ? ? 2 φ φ          n−2δ ⎠   ⎩ ? ? ? 2 φ φ φ φ φ
```

according as n has or has not a 2-digit in some j place, $j < k$. In the first case the three putative nim-values are all the jth odious number, and in the second exactly one of them is the kth odious number, so they cannot have zero nim-sum.

Now we know from our analysis of Mock Turtles that each odious number is the first number not the nim-sum of two or fewer earlier ones. It suffices to show that if n is a k-number, we can choose δ so as to make $n - \delta$ and $n - 2\delta$ i- and j-numbers or empty, for any i and j less than k. The subtraction sums in Table 6 show how to do this.

```
                 k   j   i                                           k
 δ  :            2 0 0 1 0 0         δ        1 0 0 0 1 0 1 0 0 1 0 0
 n  :     ... ? ? 2 φ φ 1 φ φ 1 φ φ  n   :    φ 2 φ φ φ 2 φ 2 φ φ 2 φ φ
n−δ :     ... ? ? ? ? ? 2 φ φ 0 φ φ  n−δ :    φ 1 φ φ φ 1 φ 1 φ φ 1 φ φ
n−2δ:     ... ? ? ? ? ? ? ? ? 2 φ φ  n−2δ:    φ 0 φ φ φ 0 φ 0 φ φ 0 φ φ

 δ  :              1 0 0 1 0 0                              k       i
 n  :     ... ? ? 2 φ φ 0 φ φ 1 φ φ  n   :    φ 2 φ φ φ 2 φ 2 φ φ 1 φ φ
n−δ :     ... ? ? ? ? ? 2 φ φ 0 φ φ  n−δ :    ? ? ? ? ? ? ? ? ? ? 2 φ φ
n−2δ:     ... ? ? ? ? ? ? ? ? 2 φ φ  n−2δ:    0 0 0 0 0 0 φ 0 0 0 φ φ

 δ  :              1 2 2 2 0 0
 n  :     ... ? ? 2 φ φ 1 φ φ 0 φ φ  n   :    φ 2 φ φ φ 2 φ 2 φ φ 0 φ φ
n−δ :     ... ? ? ? ? ? 2 φ φ 1 φ φ  n−δ :·   ? ? ? ? ? ? ? ? ? ? 2 φ φ
n−2δ:     ... ? ? ? ? ? ? ? ? 2 φ φ  n−2δ:    0 0 0 0 0 0 φ 0 0 1 φ φ

 δ  :              0 2 2 2 0 0        In the last two cases above, the first φ of
 n  :     ... ? ? 2 φ φ 0 φ φ 0 φ φ   the last line is whichever of 0 or 1 makes
n−δ :     ... ? ? ? ? ? 2 φ φ 1 φ φ   n−2δ have the same parity as n. Then δ
n−2δ:     ... ? ? ? ? ? ? ? ? 2 φ φ   can be found from n and n−2δ.
```

Table 6. How to Make $n - \delta$, $n - 2\delta$ into i- and j-numbers, or Empty.

Grunt

In a move of this game one must turn over four symmetrically arranged coins of which the first must be the leftmost coin of the game and the last must be turned from heads to tails. Numbering from 0 the restriction is that we turn numbers

$$0, a, n - a \text{ and } n, \quad 0 < a < \tfrac{1}{2}n,$$

and we find the nim-values:

n	0	1	2	3	4	5	6	7	8	9	10	11	12	13	14	15	16	17	18	19	20	21	22	...
$\mathcal{G}(n)$	0	0	0	1	0	2	1	0	2	1	0	2	1	3	2	1	3	2	4	3	0	4	3	....

Figure 8. A Winning Move in Grunt.

Since $\mathcal{G}(0) = 0$, $\mathcal{G}(n)$ can more easily be computed as the mex of all numbers of the form

$$\mathcal{G}(a) \overset{*}{+} \mathcal{G}(n - a), 0 < a < \frac{1}{2}n,$$

and so the game is a disguise for Grundy's Game (see Chapter 4, Vol. 1) in which any heap may be split into two smaller heaps of different sizes.

Sym

As an example where the nim-values display no recognizable pattern, let us turn over any symmetrically arranged set of coins, not necessarily including the leftmost coin, number 0. We find, with thanks to Donald Knuth for the last four values

$$n = 0\ 1\ 2\ 3\ 4\ 5\ 6\ \ \ 7\ \ \ 8\ \ \ 9\ 10\ 11\ 12\ 13\ \ \ 14\ 15\ \ \ 16\ 17\ \ \ 18\ 19\ \ \ \ 20\ \dots$$
$$\mathcal{G}(n) = 1\ 2\ 4\ 3\ 6\ 7\ 8\ 16\ 18\ 25\ 32\ 11\ 64\ 31\ 128\ 10\ 256\ \ 5\ 512\ 28\ 1024\ \dots.$$

The reader can also try to solve the game **Sympler** in which the leftmost coin *is* to be included in the symmetrical set of coins turned.

Two-Dimensional Turning Games

All our one-dimensional games were played with the restriction that the rightmost coin to be turned was to be changed from heads to tails. In the two-dimensional games the corresponding requirement is that the most "south-easterly" coin which is turned must go from heads to tails. In such games we'll write $\mathcal{G}(a, b)$ for the value of a coin in row a and column b.

Acrostic Twins

We start with a very simple game. The move is to turn two coins which must either be in the same row or in the same column. The typical entry in the nim-value table is therefore the least number not appearing earlier in the same row or column, and we find Table 7. So we see that Acrostic Twins defines nim-addition:

$$\mathcal{G}(a, b) = a \overset{*}{+} b.$$

Turning Corners

This is a much more interesting game. The move is to turn over the four corners of any rectangle with horizontal and vertical sides. The nim-values can be computed using

$$\mathcal{G}(a, b) = \text{mex} \left\{ \mathcal{G}(a', b) \overset{*}{+} \mathcal{G}(a, b') \overset{*}{+} (a', b') \right\},$$

where a' and b' are any numbers respectively less than a and b (see Fig. 9). Table 8 gives values for a and b less than 16.

0	1	2	3	4	5	6	7	8	9	10	11	12	13	14	15
1	0	3	2	5	4	7	6	9	8	11	10	13	12	15	14
2	3	0	1	6	7	4	5	10	11	8	9	14	15	12	13
3	2	1	0	7	6	5	4	11	10	9	8	15	14	13	12
4	5	6	7	0	1	2	3	12	13	14	15	8	9	10	11
5	4	7	6	1	0	3	2	13	12	15	14	9	8	11	10
6	7	4	5	2	3	0	1	14	15	12	13	10	11	8	9
7	6	5	4	3	2	1	0	15	14	13	12	11	10	9	8
8	9	10	11	12	13	14	15	0	1	2	3	4	5	6	7
9	8	11	10	13	12	15	14	1	0	3	2	5	4	7	6
10	11	8	9	14	15	12	13	2	3	0	1	6	7	4	5
11	10	9	8	15	14	13	12	3	2	1	0	7	6	5	4
12	13	14	15	8	9	10	11	4	5	6	7	0	1	2	3
13	12	15	14	9	8	11	10	5	4	7	6	1	0	3	2
14	15	12	13	10	11	8	9	6	7	4	5	2	3	0	1
15	14	13	12	11	10	9	8	7	6	5	4	3	2	1	0

Table 7. How to Play Acrostic Twins.

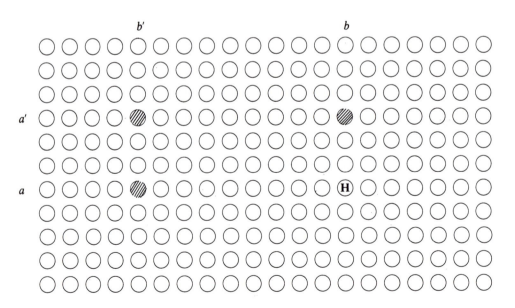

Figure 9. A Typical Move in Turning Corners.

0	0	0	0	0	0	0	0	0	0	0	0	0	0	0	0
0	1	2	3	4	5	6	7	8	9	10	11	12	13	14	15
0	2	3	1	8	10	11	9	12	14	15	13	4	6	7	5
0	3	1	2	12	15	13	14	4	7	5	6	8	11	9	10
0	4	8	12	6	2	14	10	11	15	3	7	13	9	5	1
0	5	10	15	2	7	8	13	3	6	9	12	1	4	11	14
0	6	11	13	14	8	5	3	7	1	12	10	9	15	2	4
0	7	9	14	10	13	3	4	15	8	6	1	5	2	12	11
0	8	12	4	11	3	7	15	13	5	1	9	6	14	10	2
0	9	14	7	15	6	1	8	5	12	11	2	10	3	4	13
0	10	15	5	3	9	12	6	1	11	14	4	2	8	13	7
0	11	13	6	7	12	10	1	9	2	4	15	14	5	3	8
0	12	4	8	13	1	9	5	6	10	2	14	11	7	15	3
0	13	6	11	9	4	15	2	14	3	8	5	7	10	1	12
0	14	7	9	5	11	2	12	10	4	13	3	15	1	8	6
0	15	5	10	1	14	4	11	2	13	7	8	3	12	6	9

Table 8. Have You Learnt Your Tims Table?

Nim-Multiplication

Observing that the nim-value $\mathcal{G}(0, n) = 0$, while $\mathcal{G}(1, n) = n$, we guess that this might be a kind of multiplication, so we shall write

$$a \overset{*}{\times} b$$

(and you will read "a **tims** b") for the nim-value $\mathcal{G}(a, b)$ of the general coin in Turning Corners. We shall call this the **nim-product** of a and b.

It is shown in ONAG (Chapter 6) that this remarkable operation has all the usual algebraic properties of multiplication, and in particular obeys the distributive law

$$a \overset{*}{\times} (b \overset{*}{+} c) = a \overset{*}{\times} b \overset{*}{+} a \overset{*}{\times} c$$

with nim-addition. For example

$$7 \overset{*}{\times} (5 \overset{*}{+} 6) = 7 \overset{*}{\times} 3 = 14,$$

$$7 \overset{*}{\times} 5 \overset{*}{+} 7 \overset{*}{\times} 6 = 13 \overset{*}{+} 3 = 14.$$

But note, for example, that the nim-sum of 6 and 6 is not two sixes but no sixes, since $1 \overset{*}{+} 1$ is not 2, but 0.

In computing nim-products of larger numbers, the **Fermat powers** of 2,

$$2, \quad 4, \quad 16, \quad 256, \quad 65536, \quad 4294967296, \quad \ldots, \quad 2^{2^n}, \quad \ldots$$

play a role similar to that played by *all* the powers of 2 in nim-addition. Recall that in nim-addition if N is any power of 2 we have:

$$N\overset{*}{+}n = N + n \text{ for } n < N,$$
$$N\overset{*}{+}N = 0.$$

For nim-multiplication, if N is any *Fermat* power of 2 we have:

$$N \overset{*}{\times} n = N \times n \text{ for } n < N,$$
$$N \overset{*}{\times} N = \frac{3}{2}N.$$

For example, $16 \overset{*}{\times} 5 = 80$, as usual, but $16 \overset{*}{\times} 16 = 24$. Table 9 gives products of powers of 2.

1	2	4	8	16	32	64	128	256	...
2	3	8	12	32	48	128	192	512	...
4	8	6	11	64	128	96	176	1024	...
8	12	11	13	128	192	176	208	2048	...
16	32	64	128	24	44	75	141	4096	...
32	48	128	192	44	52	141	198	8192	...
64	128	96	176	75	141	103	185	16384	...
128	192	176	208	141	198	185	222	32768	...
256	512	1024	2048	4096	8192	16384	32768	384	...

Table 9. Nim-products of Powers of 2.

Swirling Tartans

Figure 10 indicates the coins which may be turned in a typical move of this game. The boxed places form what we call a **tartan**. In general we select a certain number of rows and a certain number of columns and the places of the tartan are where our chosen rows meet our chosen columns. In **Swirling Tartans** we may turn the coins of *any* tartan, but in other games there will be restrictions on the rows and columns we may choose. Table 9 is actually a table of nim-values for Swirling Tartans. This is a particular case of the following theory.

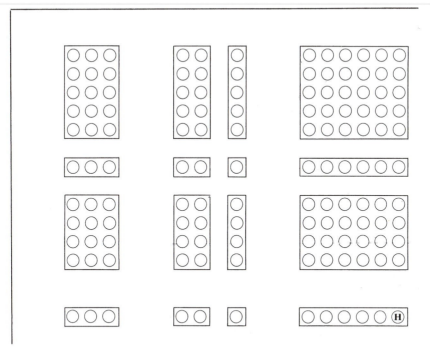

Figure 10. A Tartan.

The Tartan Theorem

We can build a **tartan game**, $A \times B$, from two one-dimensional turning games, A and B, by specifying that the rows of the tartan shall correspond to the coins which may be turned in a move of game A and the columns to the coins which may be turned in a move of game B. Taking both A and B to be the game of Motley, in which *any* sets of coins may be turned, we see that

$$\text{MOTLEY} \times \text{MOTLEY} = \text{SWIRLING TARTANS}.$$

It follows from the Tartan Theorem that the nim-values for Swirling Tartans are the nim-products of those for two games of Motley—since the latter nim-values are just the powers of 2; this justifies our assertion about Table 9. More generally:

> the nim-values for the tartan game $A \times B$ are
> the nim-products of those for A and B:
>
> $$\mathcal{G}_{A \times B}(a, b) = \mathcal{G}_A(a) \overset{*}{\times} \mathcal{G}_B(b).$$

THE TARTAN THEOREM

The proof, which we do not give, depends on the following characterizing property of the nim-product $a \overset{*}{\times} b$.

$$
\begin{array}{c}
\text{If } x_1, x_2, \ldots \text{ are numbers for which} \\
a = \operatorname{mex}(a \overset{*}{+} x_i) \\
\text{and } y_1, y_2 \ldots \text{ are numbers for which} \\
b = \operatorname{mex}(b \overset{*}{+} y_j) \\
\text{then we have} \\
a \overset{*}{\times} b = \operatorname{mex}(a \overset{*}{\times} b \overset{*}{+} x_i \overset{*}{\times} y_j).
\end{array}
$$

This can be deduced from a result on p. 55 of ONAG.

Rugs, Carpets, Windows and Doors

In Turning Corners we turned over the corners of a rectangle, so that

$$\text{TURNING CORNERS} = \text{TWINS} \times \text{TWINS}.$$

In **Rugs** we turn over *all* the coins in some solid rectangle, in other words the tartan must be defined by a block of consecutive rows and a block of consecutive columns. Since in the Ruler Game a move was to turn over a block of consecutive coins, we have

$$\text{RULER} \times \text{RULER} = \text{RUGS},$$

and so the nim-values are those of Table 10.

1	2	1	4	1	2	1	8	1	2	1	4	1	2	1	16
2	3	2	8	2	3	2	12	2	3	2	8	2	3	2	32
1	2	1	4	1	2	1	8	1	2	1	4	1	2	1	16
4	8	4	6	4	8	4	11	4	8	4	6	4	8	4	64
1	2	1	4	1	2	1	8	1	2	1	4	1	2	1	16
2	3	2	8	2	3	2	12	2	3	2	8	2	3	2	32
1	2	1	4	1	2	1	8	1	2	1	4	1	2	1	16
8	12	8	11	8	12	8	13	8	12	8	11	8	12	8	128

Table 10. A Rug with a Table on It.

A **carpet** is a tartan in which both rows and columns form symmetrical sets, as in Fig. 11, and the corresponding game, **Carpets,** has therefore the nim-values of Table 11:

$$\text{CARPETS} = \text{SYM} \times \text{SYM}.$$

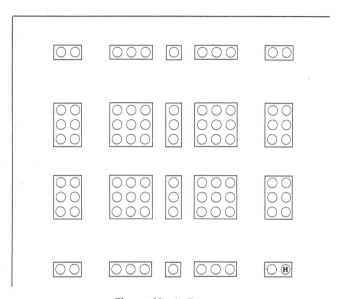

Figure 11. A Carpet.

In **Fitted Carpets** one is only allowed to turn carpets which fit snugly into the corner of the room, so

$$\text{FITTED CARPETS} = \text{SYMPLER} \times \text{SYMPLER}$$

whose analysis is left to the reader.

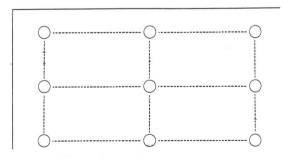

Figure 12. A Move in Windows.

Table 11. A Greatly-Valued Carpet.

In the game **Windows** we turn the nine coins where three equally spaced rows meet three equally spaced columns, as in Fig. 12, so that

$$\text{WINDOWS} = \text{TURNIPS} \times \text{TURNIPS}.$$

The nim-values form the most complex system we have yet discovered. To calculate the outcome of a given position we must perform no fewer than four successive operations:

1. Expand the two coordinates of a head in base 3 and find the last 2-digit (if any) in each.

2. Replace the coordinates by the corresponding odious numbers (or zero). This involves a further expansion, in base 2.

3. For each head find the *nim-product* of the two numbers so obtained.

4. Find the *nim-sum* of the numbers so found for all the heads.

In all these games there has been the condition that the coin most to the South-East in any move be turned from heads to tails. So that our next game deserves its name we will play it "upside-down" and impose that condition on the most North-East coin. The move in **Doors** is to turn over the twelve coins where any three equally spaced columns meet four symmetrically arranged rows, which must include the bottom row, as in Fig. 13. This shows that

$$\text{DOORS} = \text{TURNIPS} \times \text{GRUNT}$$

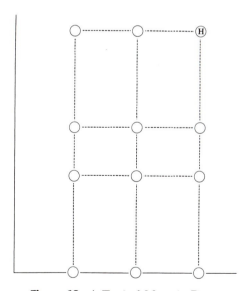

Figure 13. A Typical Move in Doors.

so you should soon be able to find the nim-value of the Doors position in which there is a single head in the 100th row and 100th column. (Beware: the first row is row number 0.)

Acrostic Games

There is another way to build a two-dimensional game out of two one-dimensional turning games, A and B. In the **acrostic product** $A \cup B$, the coins we turn must either all be in the same column or all be in the same row. If the coins are all in a column, they must

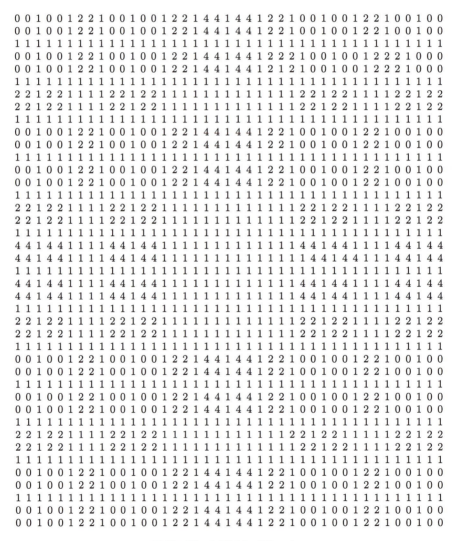

Table 12. A Field of Turnips.

correspond to a move of game A; if all in a row, to a move of game B. We have already met one game of this type:

$$\text{ACROSTIC TWINS} = \text{TWINS} \cup \text{TWINS}.$$

In **Acrostic Turnips** we must, of course, upturn three turnips which are equally spaced and either in the same row or in the same column, the furthest from the corner of the field being turned from top to tail.

$$\text{ACROSTIC TURNIPS} = \text{TURNIPS} \cup \text{TURNIPS}.$$

The first 1681 nim-values are displayed in Table 12. It is not hard to prove that a row or column that begins with zero repeats the nim-sequence for Turnips itself, while the values not in such a row or column are 1.

Stripping and Streaking

We have no idea how to play the general acrostic product $A \cup B$, even when the one-dimensional games A and B are fully understood. But if it just should happen, as sometimes it does, that every nim-value of a place in each of your games is either 7 or a power of 2, we can offer you some help. We first discuss two easy games of this kind.

In **Streaking** we turn over any collection (**streak**) of coins all in the same row or column, so that

$$\text{STREAKING} = \text{MOTLEY} \cup \text{MOTLEY}.$$

Since the nim-values for Motley are exactly the powers of 2, and each nim-value for Streaking is the mex of all numbers that are sums of earlier nim-values from the same row or earlier nim-values from the same column, we find Table 13.

In **Stripping**, the coins we turn must be consecutive (form a **strip**) in either a row or a column. The entries in Table 14, of nim-values for

$$\text{STRIPPING} = \text{RULER} \cup \text{RULER},$$

can apparently be obtained from entries in Table 13. How do we explain this?

Uglification and Derision

We introduce an ambitious distraction. We shall call an entry in Table 13 (or 14) the **ugly product** of the two powers of 2 that head its row and column, and write

$$4 \overset{*}{\cup} 8 = 10 \ (\text{"four \textbf{uggles} eight is ten"})$$

for example. Ugly products of other numbers can then be found using the distributive law:

$$4 \overset{*}{\cup} 11 = 4 \overset{*}{\cup} (8 \overset{*}{+} 2 \overset{*}{+} 1) = 10 \overset{*}{+} 5 \overset{*}{+} 4 = 11,$$
$$5 \overset{*}{\cup} 11 = (4 \overset{*}{+} 1) \overset{*}{\cup} 11 = 11 \overset{*}{+} 11 = 0.$$

The latter equation shows that 5 and 11 are **deriders of zero**. An uglification table up to 16 is given in Table 15.

1	2	4	8	16	32	64	128	256	...
2	1	5	9	17	33	65	129	257	...
4	5	2	10	18	34	66	130	258	...
8	9	10	4	20	36	68	132	260	...
16	17	18	20	8	40	72	136	264	...
32	33	34	36	40	16	80	144	272	...
64	65	66	68	72	80	32	160	288	...
128	129	130	132	136	144	160	64	320	...
256	257	258	260	264	272	288	320	128	...

Table 13. Streaking Values.

1	2	1	4	1	2	1	8	1	2	1	4	1	2	1	16
2	1	2	5	2	1	2	9	2	1	2	5	2	1	2	17
1	2	1	4	1	2	1	8	1	2	1	4	1	2	1	16
4	5	4	2	4	5	4	10	4	5	4	2	4	5	4	18
1	2	1	4	1	2	1	8	1	2	1	4	1	2	1	16
2	1	2	5	2	1	2	9	2	1	2	5	2	1	2	17
1	2	1	4	1	2	1	8	1	2	1	4	1	2	1	16
8	9	8	10	8	9	8	4	8	9	8	10	8	9	8	20
1	2	1	4	1	2	1	8	1	2	1	4	1	2	1	16
2	1	2	5	2	1	2	9	2	1	2	5	2	1	2	17
1	2	1	4	1	2	1	8	1	2	1	4	1	2	1	16
4	5	4	2	4	5	4	10	4	5	4	2	4	5	4	18
1	2	1	4	1	2	1	8	1	2	1	4	1	2	1	16
2	1	2	5	2	1	2	9	2	1	2	5	2	1	2	17
1	2	1	4	1	2	1	8	1	2	1	4	1	2	1	16
16	17	16	18	16	17	16	20	16	17	16	18	16	17	16	8

Table 14. Stripping Values.

0	0	0	0	0	0	0	0	0	0	0	0	0	0	0	0	0
0	1	2	3	4	5	6	7	8	9	10	11	12	13	14	15	16
0	2	1	3	5	7	4	6	9	11	8	10	12	14	13	15	17
0	3	3	0	1	2	2	1	1	2	2	1	0	3	3	0	1
0	4	5	1	2	6	7	3	10	14	15	11	8	12	13	9	18
0	5	7	2	6	3	1	4	2	7	5	0	4	1	3	6	2
0	6	4	2	7	1	3	5	3	5	7	1	4	2	0	6	3
0	7	6	1	3	4	5	2	11	12	13	10	8	15	14	9	19
0	8	9	1	10	2	3	11	4	12	13	5	14	6	7	15	20
0	9	11	2	14	7	5	12	12	5	7	14	2	11	9	0	4
0	10	8	2	15	5	7	13	13	7	5	15	2	8	10	0	5
0	11	10	1	11	0	1	10	5	14	15	4	14	5	4	15	21
0	12	12	0	8	4	4	8	14	2	2	14	6	10	10	6	6
0	13	14	3	12	1	1	15	6	11	8	5	10	7	4	9	22
0	14	13	3	13	3	0	14	7	9	10	4	10	4	7	9	23
0	15	15	0	9	6	6	9	15	0	0	15	6	9	9	6	7
0	16	17	1	18	2	3	19	20	4	5	21	6	22	23	7	8

Table 15. The Uglification Table up to 16's.

For larger numbers we may use the following rules:

$$
N \overset{*}{\cup} n = \begin{cases} N + \left\lfloor \dfrac{1}{2}n \right\rfloor & \text{if } n \text{ is odious} \\[2mm] \left\lfloor \dfrac{1}{2}n \right\rfloor & \text{if } n \text{ is evil,} \end{cases}
$$

If N is any power of 2 and $n < N$, we have

$$
N \overset{*}{\cup} N = \left\lceil \dfrac{1}{2}N \right\rceil .
$$

The rows and columns of Table 15 which correspond to 7 and powers of 2 have been printed in bold type and their intersections have been boxed. We shall use the following properties of these rows.

1. The entries in any **bold** row are distinct. In symbols, if a is 7 or a power of 2, and $b \neq \bar{b}$, then $a \overset{*}{\cup} b \neq a \overset{*}{\cup} \bar{b}$.

2. Each boxed entry is the mex of all previous entries in its row and column. That is, if both a and b are 7 or powers of 2, then $a \overset{*}{\cup} b$ is the mex of all numbers of the form $a' \overset{*}{\cup} b$ or $a \overset{*}{\cup} b'$, $a' < a$, $b' < b$.

These are used in proving the following theorem.

If every nim-value of a place in each of A and B is 7 or a power of 2, then the nim-values of the acrostic game $A \cup B$ are obtained by uglification of those for A and B:

$$\mathcal{G}_{A \cup B}(a, b) = \mathcal{G}_A(a) \overset{*}{\cup} \mathcal{G}_B(b).$$

THE UGLIFICATION THEOREM

To see this, let the typical move in game A or B turn the coins in places

$$a_1 < a_2 < \ldots < a$$

or

$$b_1 < b_2 < \ldots < b$$

respectively. We will denote the nim-values of these places by

$$\alpha_1, \alpha_2, \ldots, \alpha$$

and

$$\beta_1, \beta_2, \ldots, \beta$$

respectively. Then the nim-value $\mathcal{G}(a, b)$ in the acrostic product $A \cup B$ is the mex of all numbers of the form

$$\alpha_1 \overset{*}{\cup} \beta \overset{*}{+} \alpha_2 \overset{*}{\cup} \beta \overset{*}{+} \ldots$$

or

$$\alpha \overset{*}{\cup} \beta_1 \overset{*}{+} \alpha \overset{*}{\cup} \beta_2 \overset{*}{+} \ldots,$$

that is to say the mex of all numbers of the form

$$\bar{\alpha} \overset{*}{\cup} \beta \quad \text{or} \quad \alpha \overset{*}{\cup} \bar{\beta},$$

where

$$\bar{\alpha} = \alpha_1 \overset{*}{+} \alpha_2 \overset{*}{+} \dots$$
$$\bar{\beta} = \beta_1 \overset{*}{+} \beta_2 \overset{*}{+} \dots .$$

Now α is the mex of all numbers $\bar{\alpha}$ which arise in this way, and β is the mex of all numbers $\bar{\beta}$, so that every $\alpha' < \alpha$ is one of the numbers $\bar{\alpha}$ and every $\beta' < \beta$ is a $\bar{\beta}$. But each of α and β is either 7 or a power of 2 by assumption, so that certainly $\alpha \overset{*}{\cup} \beta$ is the mex of all numbers of the form

$$\alpha' \overset{*}{\cup} \beta \text{ or } \alpha \overset{*}{\cup} \beta'.$$

Also, α is distinct from all the numbers $\bar{\alpha}$, and so $\alpha \overset{*}{\cup} \beta$ is distinct from all the numbers $\bar{\alpha} \overset{*}{\cup} \beta$, and similarly from $\alpha \overset{*}{\cup} \bar{\beta}$, and so $\alpha \overset{*}{\cup} \beta$ is *their* mex also.

This explains the values we found for Stripping and enables us to discuss a few similar games. For instance, in **Strip and Streak**, where we may turn coins in a horizontal strip or a vertical streak, the first few nim-values are as in Table 16.

1	2	1	4	1	2	1	8	1	2	1	4	1	2	1	16	1	2 ...
2	1	2	5	2	1	2	9	2	1	2	5	2	1	2	17	2	1 ...
4	5	4	2	4	5	4	10	4	5	4	2	4	5	4	18	4	5 ...
8	9	8	10	8	9	8	4	8	9	8	10	8	9	8	20	8	9 ...
16	17	16	18	16	17	16	20	16	17	16	18	16	17	16	8	16	17 ...

Table 16. Strip and Streak.

Table 17 gives the nim-values for **Acrostic Mock Turtle Fives** in which we *turn up to three* coins provided that these are all contained in some horizontal or vertical strip of *five*.

1	2	4	7	8	1	2	4	7	8	1	2	4	7	8 ...
2	1	5	6	9	2	1	5	6	9	2	1	5	6	9 ...
4	5	2	3	10	4	5	2	3	10	4	5	2	3	10 ...
7	6	3	2	11	7	6	3	2	11	7	6	3	2	11 ...
8	9	10	11	4	8	9	10	11	4	8	9	10	11	4 ...
1	2	4	7	8	1	2	4	7	8	1	2	4	7	8 ...
2	1	5	6	9	2	1	5	6	9	2	1	5	6	9 ...

Table 17. Acrostic Mock Turtle Fives.

Extras

Unlocking Doors

For Turnips, Table 5 shows that $\mathcal{G}(99) = 4$, while for Grunt, the discussion of Grundy's Game in Chapter 4 shows that $\mathcal{G}(99) = 5$; so a coin in the 100th row and 100th column of Doors has value

$$4 \text{ tims } 5 = 2.$$

Sparring, Boxing and Fencing

Turning games can be played in any number of dimensions. We mention just three 3-dimensional games. In **Sparring** we turn over any two coins in the same row, column or vertical, that is to say the two ends of a "spar." The typical entry in the nim-value table is the mex of previous entries in its row, column or vertical, so we have

$$\mathcal{G}(a, b, c) = a \overset{*}{+} b \overset{*}{+} c.$$

In **Boxing** we turn the eight corners of a rectangular "box". This is the 3-dimensional version of Turning Corners and its nim-values are three-term nim-products

$$\mathcal{G}(a, b, c) = a \overset{*}{\times} b \overset{*}{\times} c.$$

In **Fencing** we turn the four corners of a rectangular "fence" whose edges are parallel to any two of the three coordinate axes. It can be shown that

$$\mathcal{G}(a, b, c) = b \overset{*}{\times} c \overset{*}{+} c \overset{*}{\times} a \overset{*}{+} a \overset{*}{\times} b.$$

In each case the furthest turned coin from the origin must go from heads to tails.

"Coins" (or Heaps) with Infinitely Many (or 2^{2^N}) "Sides"

"Coins" (or Heaps) with Infinitely Many (or 2^{2^N}) "Sides" may be used to give lots of new "turning" games whose theory also involves Nim-multiplication. Thus if the move is to alter at most two of the heaps $H_{-1}, H_0, H_1 \ldots$ of which the rightmost one must be reduced, the $\mathcal{P}$-positions are those with

$$H_0 \overset{*}{+} H_1 \overset{*}{+} H_2 \overset{*}{+} \ldots = 0 \text{ and } 0 \overset{*}{\times} H_0 \overset{*}{+} 1 \overset{*}{\times} H_1 \overset{*}{+} 2 \overset{*}{\times} H_2 \overset{*}{+} \ldots = H_{-1}$$

References and Further Reading

J. H. Conway, *On Numbers and Games* A K Peters, Ltd., Natick, MA 2001, Chapter 6.

J. H. Conway, Integral lexicographic codes, *Discrete Math.*, **73**(1990) 219–235.

J. H. Conway & N. J. A. Sloane, Lexicographic codes: error-correcting codes from game theory, *IEEE Trans. Inform. Theory*, **IT-32**(1986) 337–348.

Richard K. Guy, She loves me, she loves me not; relatives of two games of Lenstra, Een Pak met een Korte Broek, papers presented to H. W. Lenstra, 77:05:18, Mathematisch Centrum, Amsterdam.

Aviezri Fraenkel, Error-correcting codes derived from combinatorial games, in Richard Nowakowski (ed.) *More Games of No Chance*, (Berkeley CA 2000) *Math. Sci. Res. Inst. Publ.*, **42**(2002) Cambridge Univ. Press, Cambridge, UK, 417–431.

Donald Knuth, *The TeXbook*, Addison-Wesley, Reading, MA, 1984, p.241.

H. W. Lenstra, Nim multiplication, Séminaire de Théorie des Nombres, 1977–78 exposé No. 11, Université de Bordeaux.

Vera Pless, Games and codes, in R. K. Guy (ed.) Combinatorial Games, *Proc. Symp. Appl. Math.*, **43**(1991) 101–110.

-15-

Chips and Strips

There is some ill a-brewing towards my rest
For I did dream of money-bags tonight.
William Shakespeare, *The Merchant of Venice*, II, v, 17.

Many of the games in this chapter are derived in some way from Nim. Although Nim is usually played with heaps of chips it can also be played with coins on a strip, the move being to shift any coin leftwards any number of squares. Figure 1 shows the same Nim position in both versions. Moving a coin leftwards corresponds to reducing a heap.

We obtain many generalizations of Nim by varying the conditions under which heaps can be reduced, or coins moved.

Figure 1. Two Forms of Nim.

The Silver Dollar Game

In our first variant we allow at most one coin per square and do *not* allow one coin to jump over another. It can take quite a long time to discover that this is a cunning disguise for Nim, related to the game of Poker-Nim in Chapter 3. The sizes of the Nim-heaps are the lengths of *alternate* gaps between the coins starting from the rightmost coin (Fig. 2).

Observe that any *decrease* of one of these numbers is possible (by moving the coin at the right end of the gap) and that some *increases* are also possible (by moving the coin at the *left* end of a gap). We've indicated sample moves of both types in the figure. But just as in the theory of Poker-Nim, the increasing moves are mere reversible delaying moves and the winner wins by playing Nim.

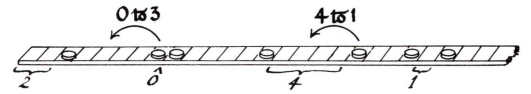

Figure 2. The Silver Dollar Game Without the Dollar.

N.G. De Bruijn has made the game more interesting by turning the leftmost square into a moneybag capable of holding any number of coins and making one of the coins a Silver Dollar, more valuable than all the others put together. Now the leftmost coin not already in the moneybag, may be put into the moneybag, as a move. The person who bags the dollar loses the game, because we also allow another move—pocket the moneybag!

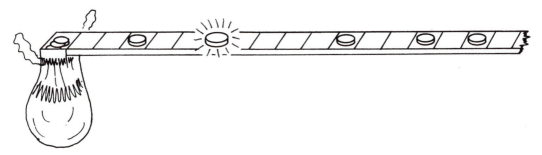

Figure 3. De Bruijn's Silver Dollar Game.

In this version the moneybag counts as a *full* square when the first coin to the right of it is the Silver Dollar; otherwise as an *empty* one (it's because we don't want to put the dollar into the bag that we think of it as full when the dollar is the nearest coin to it!) If we win the Nim game we won't be forced to put the dollar into the bag.

If we allow whoever bags the dollar to pocket the bag all in one move, we count the bag as *full* only when there's just one other coin between it and the dollar. (We don't want to put *this* coin in the bag because our opponent will make sure it is immediately followed by the dollar!)

Find the winning moves in Fig. 3 in both versions.

Profit from Gaming Tables

What do you do when you meet a game that's *not* analyzed in *Winning Ways*? You might be very lucky and get the hang of it after your first few games, but if you can't quite see what's going on and our theories don't seem to provide much of a clue, the best thing to do is to compile a **gaming table**. To do this profitably can take some skill in organizing the information. There'll be some varied examples in this chapter.

Antonim

Antipathetic Nim is Nim in which no two heaps are allowed to have the same number of chips. Of course, we don't notice empty heaps, so if you want to play it with coins on a strip the condition is that no two coins may be on the same square unless this square is the moneybag (square 0).

We can analyze 3-coin Antonim in a single table (Table 1). The headings are the sizes of two of the heaps and the entry is the unique size of heap that completes these to a $\mathcal{P}$-position. The typical entry is filled in as the least number not coinciding with any earlier entry in the same row or column, nor coinciding with either the row or column heading. An X denotes an illegal position. There is an obvious pattern showing that:

> (a, b, c) is a $\mathcal{P}$-position in Antonim just when
> $(a+1, b+1, c+1)$ is a $\mathcal{P}$-position in Nim.

	0	1	2	3	4	5	6	7	8	9	10	11	12	13	14
0	0	2	1	4	3	6	5	8	7	10	9	12	11	14	13
1	2	X	0	5	6	3	4	9	10	7	8	13	14	11	12
2	1	0	X	6	5	4	3	10	9	8	7	14	13	12	11
3	4	5	6	X	0	1	2	11	12	13	14	7	8	9	10
4	3	6	5	0	X	2	1	12	11	14	13	8	7	10	9
5	6	3	4	1	2	X	0	13	14	11	12	9	10	7	8
6	5	4	3	2	1	0	X	14	13	12	11	10	9	8	7
7	8	9	10	11	12	13	14	X	0	1	2	3	4	5	6
8	7	10	9	12	11	14	13	0	X	2	1	4	3	6	5
9	10	7	8	13	14	11	12	1	2	X	0	5	6	3	4
10	9	8	7	14	13	12	11	2	1	0	X	6	5	4	3
11	12	13	14	7	8	9	10	3	4	5	6	X	0	1	2
12	11	14	13	8	7	10	9	4	3	6	5	0	X	2	1
13	14	11	12	9	10	7	8	5	6	3	4	1	2	X	0
14	13	12	11	10	9	8	7	6	5	4	3	2	1	0	X

Table 1. $\mathcal{P}$-positions for Three-Coin Antonim.

For Antonim with 4 coins we need a 3-D table, which we can build in layers. The next layer (Table 2) suggests that there is unlikely to be a simple rule, even for positions

$$(1, a, b, c),$$

so we have cut short the further layers (Table 3).

0	X	1	2	3	4	5	6	7	8	9	10	11	12	13	14
X	2	X	0	5	6	3	4	9	10	7	8	13	14	11	12
1	X	X	X	X	X	X	X	X	X	X	X	X	X	X	X
2	0	X	X	4	3	6	5	8	7	10	9	12	11	14	13
3	5	X	4	X	2	0	7	6	9	8	11	10	13	12	15
4	6	X	3	2	X	7	0	5	12	11	14	9	8	15	10
5	3	X	6	0	7	X	2	4	11	12	13	8	9	10	16
6	4	X	5	7	0	2	X	3	14	13	12	15	10	9	8
7	9	X	8	6	5	4	3	X	2	0	15	14	16	17	11
8	10	X	7	9	12	11	14	2	X	3	0	5	4	16	6
9	7	X	10	8	11	12	13	0	3	X	2	4	5	6	17
10	8	X	9	11	14	13	12	15	0	2	X	3	6	5	4
11	13	X	12	10	9	8	15	14	5	4	3	X	2	0	7
12	14	X	11	13	8	9	10	16	4	5	6	2	X	3	0
13	11	X	14	12	15	10	9	17	16	6	5	0	3	X	2
14	12	X	13	15	10	16	8	11	6	17	4	7	0	2	X

Table 2. $\mathcal{P}$-positions $(1, a, b, c)$ for Antonim .

It's not hard to show that the $\mathcal{P}$-positions with numbers ≤ 7 are just

$$
\begin{array}{ccccccc}
(0)12, & (0)34, & (0)56, & 135, & 146, & 236, & 245, \\
1234, & 1256, & 1367, & 1457, & 2357, & 2467, & 3456 \\
& & & (0)123456. & & &
\end{array}
$$

The 3-heap $\mathcal{P}$-positions are the lines of Fig. 4, counting the top node as 0; the 4-heap ones are the complements of lines, counting it as 7.

Synonim

In **Sympathetic Nim** all heaps of the same size must be treated alike—if you reduce one heap of a given size you must reduce all heaps of that size and by the same amount (no move may affect heaps of different sizes). In the strip version the move is to take *all* the coins from some square and put them on to any earlier square.

2

0	X	1	2	3	4	5	6	7
X	1	0	X	6	5	4	3	10
1	0	X	X	4	3	6	5	8
2	X	X	X	X	X	X	X	X
3	6	4	X	X	1	7	0	5
4	5	3	X	1	X	0	7	6
5	4	6	X	7	0	X	1	3
6	3	5	X	0	7	1	X	4
7	10	8	X	5	6	3	4	X

3

0	X	1	2	3	4	5	6	7
X	4	5	6	X	0	1	2	11
1	5	X	4	X	2	0	7	6
2	6	4	X	X	1	7	0	5
3	X	X	X	X	X	X	X	X
4	0	2	1	X	X	6	5	8
5	1	0	7	X	6	X	4	2
6	2	7	0	X	5	4	X	1
7	11	6	5	X	8	2	1	X

4

0	X	1	2	3	4	5	6	7
X	3	6	5	0	X	2	1	12
1	6	X	3	2	X	7	0	5
2	5	3	X	1	X	0	7	6
3	0	2	1	X	X	6	5	8
4	X	X	X	X	X	X	X	X
5	2	7	0	6	X	X	3	1
6	1	0	7	5	X	3	X	2
7	12	5	6	8	X	1	2	X

5

0	X	1	2	3	4	5	6	7
X	6	3	4	1	2	X	0	13
1	3	X	6	0	7	X	2	4
2	4	6	X	7	0	X	1	3
3	1	0	7	X	6	X	4	2
4	2	7	0	6	X	X	3	1
5	X	X	X	X	X	X	X	X
6	0	2	1	4	3	X	X	8
7	13	4	3	2	1	X	8	X

Table 3. $\mathcal{P}$-positions (k, a, b, c) for Antonim, $2 \le k \le 5$.

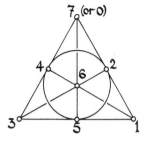

Figure 4. Fano's Fancy Antonim Finder.

This game need not detain us long. Since all the heaps of a given size must be treated in the same way they may be regarded as a single heap. A move reducing this heap to the size of an already existing heap has the same effect as removing the heap entirely. We might as well say that the heaps must always be of different sizes, so

---

SYNONIM is
just a synonym
for ANTONIM!

Simonim

SImilar MOve NIM, or **Simonim**, was rediscovered by Simon Norton. It is just Nim with the additional feature that a player may make any number of moves provided that these are all exactly similar, i.e. that they all reduce some number a to another number b. It differs from Synonim in that we are *not* required to reduce *all* the heaps of a given size. If we play it with coins on a strip the rule becomes that any number of coins may be moved from any square to any earlier square, occupied or not. Table 4 is a bit harder to compute.

	0	1	2	3	4	5	6	7	8	9	10	11	12	13
0	→↓0	2	1	4	3	6	5	8	7	10	9	12	11	14
1	2	3	0	↓1	→4	7	8	5	6	11	12	9	10	15
2	1	0	4	→3	↓2	8	7	6	5	12	11	10	9	16
3	4	→1	↓3	2	0	9	10	11	12	5	6	7	8	17
4	3	↓4	→2	0	1	10	9	12	11	6	5	8	7	18
5	6	7	8	9	10	11	0	1	2	3	4	↓5	→12	19
6	5	8	7	10	9	0	12	2	1	4	3	→11	↓6	20
7	8	5	6	11	12	1	2	9	0	↓7	→10	3	4	21
8	7	6	5	12	11	2	1	0	10	→9	↓8	4	3	22
9	10	11	12	5	6	3	4	→7	↓9	8	0	1	2	23
10	9	12	11	6	5	4	3	↓10	→8	0	7	2	1	24
11	12	9	10	7	8	→5	↓11	3	4	1	2	6	0	25
12	11	10	9	8	7	↓12	→6	4	3	2	1	0	5	26
13	14	15	16	17	18	19	20	21	22	23	24	25	26	27

Table 4. *P*-positions for Three-Coin Simonim.

As usual we try to fill in the least number not seen earlier in the row or column (or diagonal, if the entry is on the diagonal). But at most one entry n (written $\vec{n}$) in a row may equal its column label, at most one entry m (written $\downarrow m$) in a column may equal its row label and only the diagonal entry 0 may coincide with its row *and* column label.

When you've stared at Table 4 for an hour or two you'll notice various patterns which make the structure crystal clear. The solid dividing lines mark the closing up of the leading

$$1 \times 1, \ 5 \times 5, \ 13 \times 13,$$

and, in general

$$2^n - 3 \text{ by } 2^n - 3$$

portions, which form Latin squares. The arrowed entries fall into 2×2 boxes. Various portions of the table resemble the nim-addition table.

After we'd extended the table into three dimensions and enlarged it in two, we were able to work out a general rule for 4-heap Simonim. With Simon's help we were even able to prove it!

Partition the positive integers into **ranges**

$$1, \ 2 - 3, \ 4 - 7, \ 8 - 15, \ 16 - 31, \ \ldots .$$

Then transform the Simonim position as follows:
Replace the first occurrence of a number n by n',
a second occurrence by n'', and a third by n''', where

$$n' = \begin{cases} n + 3 & \text{if this is in the largest range that is} \\ & \text{represented in the transformed position,} \\ n + 1 & \text{if this is in the next largest range that is} \\ & \text{represented in the transformed position,} \\ n & \text{otherwise.} \end{cases}$$

$$n'' = \begin{cases} \text{the largest number in the range before } n', \text{ or the} \\ \text{next-to-largest number in this range,} \\ \quad \text{if } n \text{ is the next-to-largest number} \\ \quad \text{in the original position.} \end{cases}$$

$$n''' = \text{the largest number from the range before } n''$$

The original position will be a $\mathcal{P}$-position in SIMONIM just if the transformed position is a $\mathcal{P}$-position in NIM.

RULE FOR 4-HEAP SIMONIM

In applying the rule it's best to write the numbers in descending order. What should we do from

n	16	9	4	1?
n'	19	10	4	1

We find

whose nim-sum involves 16, so it can't be a $\mathcal{P}$-position. We must therefore decrease 16 to some value x for which $x + 3$ won't be in the $16 - 31$ range. Then

n	x	9	4	1
n'	?	12	5	1

so ? must be $12 \overset{*}{+} 5 \overset{*}{+} 1 = 8$. Since this is in the largest range to appear in the transformed position,

$$x \text{ must be } 8 - 3 = 5.$$

Let's do

n	9	9	7	2	
n'	12		10	2	$\Big\}$nim-sum B.
n''		7			

We can change the nim-sum to 0 by changing 2 into 1, 10 into 9 or 7 into 4, yielding the positions

n	9	9	7	1	9	9	6	2	9	3	7	2
n'	12		10	1	12		9	2	12	4	10	2
n''		7				7						

A really tricky example is

n	44	33	22	11	
n'	47	36	23	11	nim-sum 23

There's no hope of changing any of the nim-values 47, 36, or 11 by 23, while if we *remove* the 22 heap:

n	44	33	0	11	
n'	47	36	0	12	nim-sum 7

we don't arrive at a $\mathcal{P}$-position. The trick is to equalize the two small heaps:

n	44	33	11	11	
n'	47	36	12		$\Big\}$nim-sum 0.
n''				7	

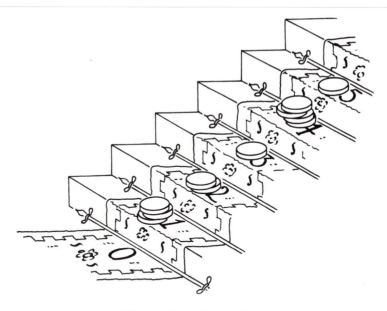

Figure 5. Stacks on Stairs.

Staircase Fives

You play this with coins on a staircase (Fig. 5). The move is to take any number, *less than five*, of coins from one step and put them on any lower step, *less than five* stairs away. The winner is the one who puts the last coin on the bottommost step.

If there are only 4 coins and 5 steps, the "five" restrictions don't matter and the game reduces to Antonim. A study of the upper 5×5 portion of Tables 1–3 provides an unexpectedly simple rule:

> Mentally interchange the coins on steps 2 and 4.
> Then the position is a $\mathcal{P}$-position just if the sum
> of the heights of all the coins is a multiple of 5.

Thus you should arrange that after your move, if there are

$$a \text{ coins on } 0, \ b \text{ on } 1, \ c \text{ on } 2, \ d \text{ on } 3 \text{ and } e \text{ on } 4,$$

then

$$0 \cdot a + 1 \cdot b + 4 \cdot c + 3 \cdot d + 2 \cdot e$$

is divisible by 5.

The rule continues to apply with more coins and more steps provided we interchange steps $5n + 2$ and $5n + 4$.

Twopins (pronounced "Tuppins")

is a bowling game which generalizes Kayles (·**77**) and Dawson's Kayles (·**07**) in Chapter 4. This time the pins are set up in columns of 1 or 2 and the condition is that, as in Kayles, a legal shot must remove just 1 or 2 adjacent columns. But there is the additional rule that it is illegal to remove just a single pin.

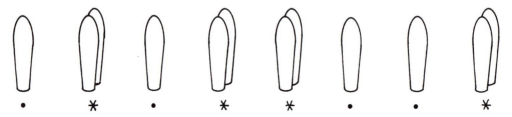

Figure 6. A Game of Twopins.

In discussing Twopins configurations we'll use

$\bullet$ for a column of one,
$*$ for a column of two,

so that the 2^n possible configurations of n non-empty Twopins columns are represented by the 2^n sequences of $n *$'s and $\cdot$'s. For instance we find

$$\bullet = 0, \quad \bullet\bullet = \bullet\bullet\bullet = \bullet*\ = *\bullet\bullet\ = *, \quad *\bullet* = *\ +\ * = 0,$$

and happily $*$, $**$, $***$ have values $*$, $*2$, $*3$; however, $****= *$. Fortunately we don't need to list all possible sequences separately because there are several useful equivalences. For example it's easy to see that

~~~~~~$*\bullet$ ~~~~~~$\bullet\bullet$ and   ~~~~~~$*$

all behave the same in play, while

~~~~~~$*\bullet\bullet*$~~~~~   behaves like   ~~~~~$*\bullet\bullet*$~~~~~~

There is also a useful **Twopins Decomposition Theorem**:

$$\sim\sim\sim *\bullet* \sim\sim\sim\ =\ \sim\sim\sim * + * \sim\sim\sim$$

After these theorems you can suppose that all strings have stars at each end and that dots come in internal blocks of three or more. Also the sequence $(*)^n$ of n stars behaves like the Kayles position K_n, while the sequences $(\cdot)^n$ and $*(\cdot)^{n-4}*$ behave like D_n in Dawson's Kayles, so you can read off their values from Chapter 4. Our Twopins-Wheel (Fig. 7) gives the nim-values of all other Twopins sequences of length 9 or less, except

$$*\bullet\bullet\bullet*\bullet\bullet\bullet*, \text{ nim-value 1}$$

Figure 7. A Twopins-Wheel.

All our equivalences remain valid in misère play, but the entries in the Twopins-Wheel should be replaced according to the scheme

For	read	genus	For	read	genus
0	0	0^1	0	$2+2$	0^0
1	1	1^0	1	$3+2$	1^1
2	2	2^2	2	$k_1k3_22_230$	2^2
3	3	3^3	3	$2_221 = d$	3^{1431}
4	$2_2321 = k$	4^{146}	4	3_2320	4^{046}
5	$k+1$	5^{057}	5	$kd3_2210$	5^{3146}

Twopins has applications to Dots-and-Boxes (Chapter 16) and to Cram.

Cram

Cram is Martin Gardner's name for impartial Domineering. It has also been called Plugg and Dots-and-Pairs, and is associated with the names of Geoffrey Mott-Smith, Sol Golomb and John Conway. You play it just like ordinary Domineering (Chapter 5; Martin Gardner called it Crosscram) except that *either* player may place his dominoes in *either* direction. You just cram them in however you can.

If you start with a rectangle with even dimensions, then there's a simple symmetry strategy for the second player if the aim is to be the last to move. If one dimension is even and the other odd, then the first player wins with the same strategy after placing her first domino at the centre of the board.

It helps to see what's going on if you replace the available regions by graphs with nodes for squares, joined by edges when they're adjacent, as we did for Col (Chapter 2) and Snort (Chapter 6).

In this form the move is to delete two adjacent nodes and all edges running up to them, and you can play the game on arbitrary graphs. Only the abstract structure of the graph matters, so that many differently shaped regions can have the same graph, e.g.

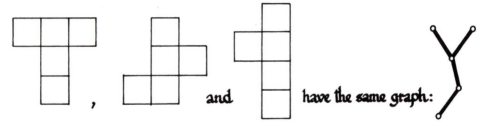

This graph, like many others, is a caterpillar. Formally, a **caterpillar** is a graph whose **body** consists of a chain of nodes, some of which may have **tufts** (or legs) i.e. 1 or more edges

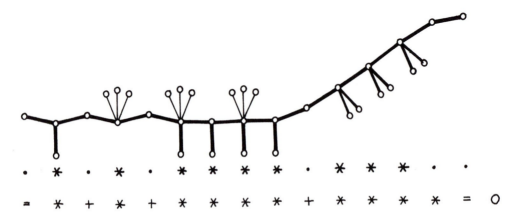

Figure 8. Even a Complicated Cram Caterpillar is a Twopins Position.

leading to otherwise isolated nodes. Luckily, even the most complicated caterpillar (Fig. 8) is equivalent to a Twopins configuration, by letting

- • replace each untufted body node, and
- ∗ represent any tufted one.

In this notation our Twopins equivalences become

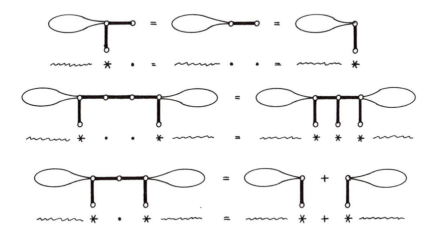

the last of which we might indicate in a single picture, using thin lines for edges which can all be omitted without affecting its value.

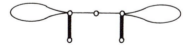

The balloons in our pictures need not be caterpillar-shaped and can even be allowed to meet, so our last identity becomes

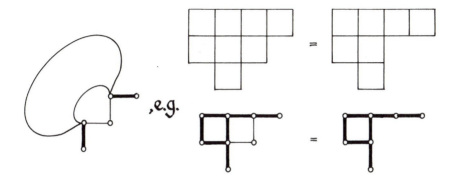

The diagrams in Fig. 9 have similar properties for example, the deletion of *all* thin edges in a diagram will not affect its value.

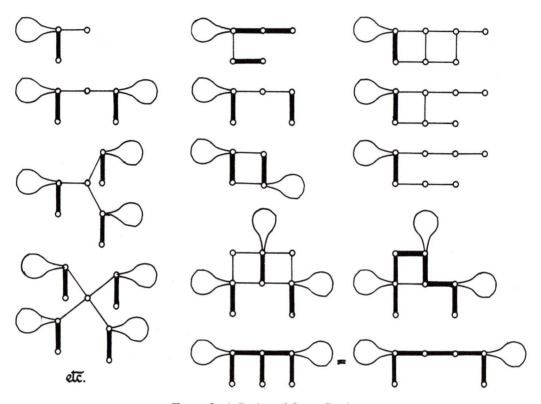

Figure 9. A Packet of Cram Crackers.

Table 5 gives values of a number of Cram positions. The dotted line indicates a chain of n edges where n is at the head of the table. We've given the full genus so that you can play Misère Cram. We use the letters

$$
\begin{array}{lll}
k & \text{for} & 2_2321 \quad \text{(position } K_5 \text{ in Kayles)}, \\
d & \text{for} & 2_221 \quad \text{(position } D_{10} \text{ in Dawson's Kayles)}, \\
e & \text{for} & 2_231 \quad \text{(arises in the } Ex\text{-Officer's Game, } \cdot\mathbf{06}\text{), and} \\
f & \text{for} & 2_21 \quad \text{(arises in Flanigan's Game, } \cdot\mathbf{34}\text{)}
\end{array}
$$

in the last column, to list the games which are not Nim-heaps.

Some other values appear in Fig. 10. The ladder values in particular are easy to remember and the remark about tufts makes them extremely useful.

number of edges in the dotted line	0	1	2	3	4	5	6	non-Nim-heaps
	1	1	2	0	3	1	1	–
	2	2	3	3	1	2	4^{146}	k
	3	3	1	2	4^{146}	3	3	k
	1	1	4^{146}	0	3	5^{057}	2^2	$k, k+1, k_1k3_22_230$
	3	3	2	2	0	3	5^{057}	$kd3_2320$
	1	1	2	0	3	1	2^{0520}	$k3_230$
	0	0	1	1	2	2^{1420}	3	f
	0	3	1	2	2^{1420}	3	5^{3146}	$e, ked3_210$
	3	1	2	0	3^{1431}	3	2^{0520}	$d, d+1$
	1	1	2	0	3	1	2^{0520}	$kd3_230$
		3	1	0^0	4^{146}	1^1	3	$2_2, k, 3_2$
		1	2	2	3	3^{31}	5^{057}	$kf3_2210, k+1$
		2	2^{1420}	0	1	1	2	f

Table 5. The Genus of Various Cram Positions.

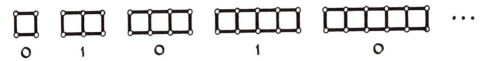

(The values of these ladders alternate and are unaffected by the addition of up to two tufts.)

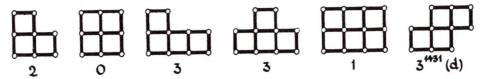

Figure 10. A Few More Cram Values.

Welter's Game

This is the coin-on-strip game in which at most one coin may be on a square, and any leftwards move of a coin onto an empty square is permitted, even if it passes over other coins. Victor Meally has observed that if overtaking is forbidden, then we have the Silver Dollar Game. Although the simplest cases were investigated by Roland Sprague, C.P. Welter discovered many remarkable properties of the general case. A simplified version of the theory is given in ONAG. Here we'll just tell you the answers and describe some new discoveries.

We'll write

$$[a|b|c|\ldots]_k \quad (\text{''}a \text{ \textbf{welt} } b \text{ welt } c \text{ welt} \ldots \text{''})$$

for the nim-value of the Welter's Game position with coins on the k different squares

$$a, b, c, \ldots,$$

and will often omit k when the number of terms is clear. The easiest way to compute this **Welter function** is the **Mating Method**.

Mate those two of the k numbers that are congruent modulo the highest power of 2. Then select a pair of mates from the remaining $k - 2$ numbers by the same rule, and so on. Eventually we have mated all except possibly one of the numbers (the **spinster**, s), say as

$$(a, b), (c, d), \ldots, \text{and possibly } s.$$

Then

$$[a|b|c|\ldots]_k = [a|b] \overset{*}{+} [c|d] \overset{*}{+} \ldots (\overset{*}{+} s \text{ if } k \text{ is odd}).$$

The two-term Welter function can be evaluated using the formula

$$[x|y] = (x \overset{*}{+} y) - 1.$$

For example,

$$[2|3|5|7|11|13|17|19]_8$$
$$= [3|19] \overset{*}{+} [5|13] \overset{*}{+} [7|11] \overset{*}{+} [2|17]$$
$$= 15 \overset{*}{+} 7 \overset{*}{+} 11 \overset{*}{+} 18 = 17,$$

there being no spinster in this case, while in

$$[0|1|4|9|16|25|36]$$
$$= [4|36] \overset{*}{+} [0|16] \overset{*}{+} [9|25] \overset{*}{+} 1$$
$$= 31 \overset{*}{+} 15 \overset{*}{+} 15 \overset{*}{+} 1 = 30$$

we see that 1 is the spinster. In this example there were two equally well mated pairs, (0,16), (9,25). In such cases it doesn't matter which pair we mate first.

Four-Coin Welter is Just Nim

When you play a few games you'll soon notice, like many other people, that a Nim-like strategy suffices for Welter's Game with four coins, so that

$$[a|b|c|d] = 0 \text{ just if } a \overset{*}{+} b \overset{*}{+} c \overset{*}{+} d = 0.$$

Welter's theory explains this by noting that if a, b and c, d are the mates, these equalities reduce to

$$[a|b] = [c|d] \text{ and } a \overset{*}{+} b = c \overset{*}{+} d$$

which are equivalent since $[x|y] = (x \overset{*}{+} y) - 1$.

And So's Three-Coin Welter!

If one of your four coins is on 0, you're really just playing three-coin Welter with the others, but shifted one place. In symbols

$$[0|a|b|c] = [a-1|b-1|c-1]$$

or

$$[a|b|c] = [0|a+1|b+1|c+1].$$

Thus the Welter position with coins on 2, 5, 7 is equivalent to the Welter or Nim position with coins on $0, 3, 6, 8$, which is cured by moving 8 to 5, so in the three-coin position we should move 7 to 4.

The Congruence Modulo 16

Although the Mating Method makes it very easy to work out nim-values, it's not so easy to find which move you should make to restore the nim-value to 0. But if the number of coins is a multiple of 4, there is a remarkable connexion with Nim:

$$[a|b|c| \ldots]_{4k} \equiv 0, \bmod 16$$
$$\text{exactly when}$$
$$a \overset{*}{+} b \overset{*}{+} c \overset{*}{+} \ldots \equiv 0, \bmod 16.$$

   This ensures in particular that when the $4k$ coins are among the first 16 places the Welter's game $\mathcal{P}$-positions are exactly the $\mathcal{P}$-positions in Nim that have distinct numbers. What are the good moves from

$$(0, 1, 2, 3, 5, 7, 11, 13)?$$

The numbers	0	1	2	3	5	7	11	13	nim-add to 4,
and we get	4	5	6	7	1	3	15	9	by nim-adding 4 to them.
But the marks	×	×	×	×	×	×	×	√	show that only the last of these is legal

(the rest involve increases or moves to occupied squares) so the only good move is from 13 to 9. Now let's look at

	2	3	5	7	11	13	17	19,	nim-sum 7,
giving	5	4	2	0	12	10	22	20,	on nim-adding 7,
which reduce to	5	4	2	0	12	10	6	4,	modulo 16.
	×	×	×		×				

So the only hopeful moves are

$$7 \text{ to } 0, 13 \text{ to } 10, 17 \text{ to } 6 \text{ and } 19 \text{ to } 4.$$

But of the Welter functions

$$[2|3|5|0|11|13|17|19], \quad [2|3|5|7|11|10|17|19], \quad [2|3|5|7|11|13|6|19], \quad [2|3|5|7|11|13|17|4]$$

only the third can be zero (glance at the mate of 17 to see that the binary expansion of the others must have a 16-digit). So the unique good move is from 17 to 6.

   What happens when the number of coins *isn't* a multiple of 4? If there are 6 coins, say, on positions

$$1, 2, 3, 5, 8, 13$$

you can pretend that there are really 8 coins on places

$$-2, -1, 1, 2, 3, 5, 8, 13$$

on a strip you've perversely numbered starting from -2. Renumbering from 0 we see the position

	0	1	3	4	5	7	10	15	nim-sum 1
yielding	1	0	2	5	4	6	11	14	
	×	×		×	×		×		

so this time there are three good moves, from

	3 to 2,	7 to 6,	15 to 14,	in the new notation,
or	1 to 0,	5 to 4,	13 to 12,	in the old.

If there are 5 coins, say on

			2	3	5	7	11	
we increase by 3:	0	1	2	5	6	8	10	14	nim-sum 12
yielding			9	10	4	6	2	
			×	×		×	×	
decreasing again:					1			

showing that the only good move is from 5 to 1.

Frieze Patterns

Patterns of numbers such as

```
1   1   1   1   1   1   1   1   1   1   1       1       1       ...
  1   2   2   3   1   2   4   1   2   2   3       1       ...
    1   3   5   2   1   7   3   1   3   5       2       ...
      1   7   3   1   3   5   2   1   7   3       ...
        1   2   4   1   2   2   3   1   2   4       ...
      1   1   1   1   1   1   1   1   1   1   ...
```

(in which each diamond of numbers

$$\begin{array}{ccc} & b & \\ a & & d \\ & c & \end{array}$$ satisfied $ad = bc + 1$ so that $d = \dfrac{bc+1}{a}$),

have many wonderful properties. For example, if you start with two horizontal rows of 1's connected by any zigzag of intermediate 1's, say

```
  1   1   1   1   1   1   1   1   1   1   1
    1   ?   ?   ?   ?   !   ?   ?   !   .   .
      1   ?   ?   ?   !   ?   ?   ?   !   .   .
        1   ?   ?   ?   !   ?   ?   ?   !   .   .
          1   ?   ?   !   ?   ?   ?   ?   !   .
            1   1   1   1   1   1   1   1   1
```

you'll find that all the entries are whole numbers and that each ! is 1 so that the pattern repeats itself alternately one way up then the other. These self-checking properties mean that your children can have fun while practising their arithmetic. If you want to check your own arithmetic on the above example, see the Extras.

G. C. Shephard has observed that we can replace multiplication by addition, making each diamond

$$\begin{array}{ccc} & b & \\ a & & d \\ & c & \end{array}$$ satisfy $(a + d) = (b + c) + 1$ so that $d = b + c + 1 - a$.

If we replace the starting 1's by 0's the resulting pattern

```
0   0   0   0   0   0   0   0   0   0   0   0   0       ...
  0   1   2   4   3   0   1   2   4   3   0   1       ...
    0   2   5   6   2   0   2   5   6   2   0   2       ...
      0   4   6   4   1   0   4   6   4   1   0       ...
        0   1   4   3   2   0   1   4   3   2   0   1       ...
      0   0   0   0   0   0   0   0   0   0   0   0   ...
```

has similar properties, but this time it repeats itself the same way up.

We thought it might be a good idea to take the basic operation as nim-addition rather than ordinary multiplication or addition, and to our great surprise found that we had discovered a new way of calculating the Welter function!

You start with a row of zeros above the Welter position you want to evaluate, and work downwards making each diamond

$$
\begin{array}{c}
b \\
a \quad\quad d \\
c
\end{array}
\quad \text{satisfy } (a \overset{*}{+} d) = (b \overset{*}{+} c) + 1 \text{ so that } c = ((a \overset{*}{+} d) - 1) \overset{*}{+} b,
$$

and, in the unlikely event that you make no mistakes, you'll find the Welter function at the bottom of the triangle; e.g.

```
 0   0   0   0      0      0       0      0    0
   2   3   5   7      11     13      17     19
     0   5   1    11      5      27      1
       7   6    14      6      16      8
         5   6    12     16      12
           4   7    29     11
             4   21      5
              23     18
                17
```

yields the same answer as before, so *we* probably haven't made any mistakes! This rule is equivalent to the identity

$$[a|b|\ldots|y|z]_{k+1} = [[a|b|\ldots|y]_k] \overset{*}{+} [b|\ldots|y]_{k-1}$$

which you can find in ONAG (p. 159).

Although by hand this calculation seems much longer, it's quite a good technique to use if you want to teach your computer to play Welter's Game.

Inverting the Welter Function

Suppose that you've evaluated

$$[a|b|c|\ldots] = n$$

and have in mind a number $n' \neq n$. Then there are unique numbers

$$a' \neq a, \ b' \neq b, \ c' \neq c, \ldots$$

for which

$$
\begin{aligned}
{[a'|b|c|\ldots]} &= n', \\
{[a|b'|c|\ldots]} &= n', \\
{[a|b|c'|\ldots]} &= n', \\
&\cdots\cdots\cdots
\end{aligned}
$$

Moreover it can be shown that the equation

$$[a|b|c|\ldots] = n$$

remains true if any *even* number of the letters $a, b, c, \ldots, n$ are replaced by the corresponding primed letters. We express this happy state of affairs by the single "equation"

$$\begin{bmatrix} a & b & c & \cdots \\ a' & b' & c' & \cdots \end{bmatrix} = \begin{matrix} n \\ n' \end{matrix}$$

Using this Even Alteration Theorem and the properties of frieze patterns your computer can invert the Welter function.

For example, if you have five numbers with Welter function

$$[a|b|c|d|e] = n,$$

and want to find the numbers

$$a', b', c', d', e'$$

for which

$$\begin{bmatrix} a & b & c & d & e \\ a' & b' & c' & d' & e' \end{bmatrix} = \begin{matrix} n \\ n' \end{matrix}$$

you should complete the frieze pattern

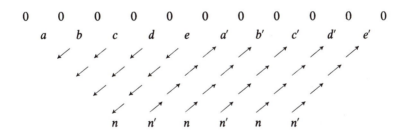

in which n and n' alternate along the bottom row, by working in the directions shown by the arrows.

Thus to find the good moves from

$$1 \quad 4 \quad 9 \quad 16 \quad 25$$

we must change one number to make the Welter function 0. The calculation

```
0   0    0    0    0    0  /  0    0    0    0    0  /  0    0
  1    4    9   16   25  /  36   33   12   13   28  /  1    4
     4   12   24    8  /  60    4   44    0   16  /  28    4
        3   26   31  /  42   19    6   39    2  /  23   22
          20   28  /  60    4   16   12   36  /  4    28
             29  /  0   29    0   29    0  /  29    0
```

(in which the rightmost two diagonals are only for checking) shows that

$$\left[\begin{array}{c|c|c|c|c} 1 & 4 & 9 & 16 & 25 \\ 36 & 33 & 12 & 13 & 28 \end{array} \right] = \begin{array}{c} 29 \\ 0 \end{array}$$

and so the only good move is from 16 to 13.

The Abacus Positions

One day we idly wrote down the infinite frieze pattern

```
...  0   0   0   0   0   0   0   0 / 0 \ 0   0   0   0   0   0   0   0  ...
... 14  12  10   8   6   4   2 / 0   1 \ 3   5   7   9  11  13  15  ...
...  1   5   1  13   1   5 / 1   0   1 \ 5   1  13   1   5   1  ...
... 15   9   3  13   7 / 1   0   1   0 \ 6  12   2   8  14  ...
...  0   8   0   8 / 0   1   0   1   0 \ 8   0   8   0  ...
... 14   4  10 / 0   1   0   1   0   1 \11   5  15  ...
...  1  13 / 1   0   1   0   1   0   1 \13   1  ...
... 15 / 1   0   1   0   1   0   1   0 \14  ...
... / 0   1   0   1   0   1   0   1   0 \ ...
  / ...  ...  ...  ...  ...  ...  ...  ...  ...
```

which suggested to us the sequence of equations

$$\left[\begin{array}{c} 0 \\ 1 \end{array} \right] = \begin{array}{c} 0 \\ 1 \end{array} \quad \left[\begin{array}{c|c} 2 & 0 \\ 1 & 3 \end{array} \right] = \begin{array}{c} 1 \\ 0 \end{array} \quad \left[\begin{array}{c|c|c} 4 & 2 & 0 \\ 1 & 3 & 5 \end{array} \right] = \begin{array}{c} 1 \\ 0 \end{array} \quad \left[\begin{array}{c|c|c|c} 6 & 4 & 2 & 0 \\ 1 & 3 & 5 & 7 \end{array} \right] = \begin{array}{c} 0 \\ 1 \end{array} \cdots$$

$$\cdots \left[\begin{array}{c|c|c|c|c|c|c|c} 14 & 12 & 10 & 8 & 6 & 4 & 2 & 0 \\ 1 & 3 & 5 & 7 & 9 & 11 & 13 & 15 \end{array} \right] = \begin{array}{c} 0 \\ 1 \end{array} \cdots$$

Since we can interchange any even number of the pairs

$$(a, a'), (b, b'), (c, c'), \ldots, (n, n'),$$

we can reorder these equations to say

$$\begin{bmatrix} 0 \\ 1 \end{bmatrix} = \begin{bmatrix} 0 & 1 \\ 3 & 2 \end{bmatrix} = \begin{bmatrix} 0 & 1 & 2 \\ 5 & 4 & 3 \end{bmatrix} = \begin{bmatrix} 0 & 1 & 2 & 3 \\ 7 & 6 & 5 & 4 \end{bmatrix} = \begin{bmatrix} 0 & 1 & 2 & 3 & 4 \\ 9 & 8 & 7 & 6 & 5 \end{bmatrix} = \dots = \begin{matrix} 0 \\ 1 \end{matrix}$$

For some reason the particular equation

$$\begin{bmatrix} 0 & 1 & 2 & 3 & 4 \\ 9 & 8 & 7 & 6 & 5 \end{bmatrix} = \begin{matrix} 0 \\ 1 \end{matrix}$$

made us think of our abacus (Fig. 11) so we call the positions evaluated in the equation

$$\begin{bmatrix} 0 & 1 & 2 & \dots & k-3 & k-2 & k-1 \\ 2k-1 & 2k-2 & 2k-3 & \dots & k+2 & k+1 & k \end{bmatrix}_k = \begin{matrix} 0 \\ 1 \end{matrix}$$

the k-coin **Abacus Positions**. Thus the equation

$$[9|1|2|6|5]_5 = 1$$

shows a 5-coin Abacus Position with its Welter function (or nim-value) 1.

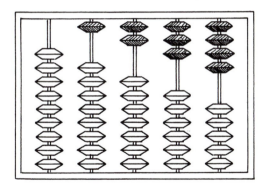

Figure 11. Swanpan.

The Abacus Strategy

We can give an explicit strategy for the Abacus Positions. Let the putative equation
$$[a|b|c|\dots]_k = 0$$
represent one of the Abacus Positions which we believe has Welter function 0. Define

$$a' = 2k - 1 - a, \quad b' = 2k - 1 - b, \quad c' = 2k - 1 - c, \quad \dots$$

and note that we have an even number of

$$a > a', \ b > b', \ c > c', \ \dots.$$

Now suppose your opponent makes the move which replaces a by x. Then since every number $\leq 2k - 1$ appears in the list

$$a, a', b, b', c, c', \dots,$$

x must be one of
$$a', b', c', \ldots,$$
say b' or a'. If $x = b'$ we must have
$$a > b', \text{ and so } b > a'$$
and we can respond with the move from b to a' since
$$[b'|a'|c\ldots]_k = 0$$
represents a simpler Abacus Position. If $x = a'$, then we have $a > a'$ and therefore an *odd* number of
$$b > b', \ c > c', \ \ldots$$
so that we can respond with one of the moves
$$b \text{ to } b', \ c \text{ to } c', \ldots.$$
A similar strategy shows that if
$$[a|b|c|\ldots]_k = 1$$
represents one of the Abacus Positions asserted to have Welter function 1, then for every move our opponent makes except the very last move of the game, we can reply with a move to another such position.

The Misère Form of Welter's Game

The remarks we've just made show not only that the Abacus Positions really do have the asserted nim-values 0 and 1, but actually that they are equivalent to nim-heaps of sizes 0 and 1 even in the misère form of Welter's Game, for it is easy to see that there is a move from any non-terminal Abacus Position to an Abacus Position of the other value. In the language of Chapter 13,

> every Abacus Position is *fickle,*

because nim-heaps of sizes 0 and 1 swap outcomes when we change from normal to misère play. On the other hand,

> every *non*-Abacus Position is *firm,*

a result which establishes that

> Welter's Game is really tame!

It suffices to show that if we can move from some *non*-Abacus Position

$$(x, b, c, \ldots) \ldots$$

to some Abacus Position

$$(a, b, c, \ldots)$$

then we could also have moved to an Abacus Position of the opposite value. But in our previous notation, if $x > a'$ we can move to

$$(a', b, c, \ldots).$$

Otherwise we must have $x < a'$, since

$$(a', b, c, \ldots)$$

is an Abacus Position. Since all numbers $\leq 2k - 1$ appear among

$$a, a', b, b', c, c', \ldots$$

we can suppose that $x = b'$, say, whence

$$b' < a' \text{ and so } a < b$$

so that we could have moved to

$$(b', a, c, \ldots).$$

> If you intend to *lose* Welter's Game, play as if you meant to win, until this would make you move into an Abacus Position, and then move instead to an Abacus Position of the opposite kind.

T. H. O'Beirne considered the misère form of Welter's Game. However our complete analysis, which independently reaches the same conclusions as that of Yamasaki, shows that his simple rule only works for very small numbers.

Kotzig's Nim

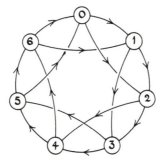

Figure 12. Kotzig's Nim on a 7-Place Strip with Move Set {1,2}.

You play this by placing coins on a circular strip. Start by placing a coin on any square—after that each player in turn puts a coin just m places further round the strip in a clockwise direction from the last coin placed. You must choose m from a previously decided **move set**. You lose if all the places where you might put coins are already occupied—you're only allowed to put one coin in any one place. Figure 12 is the directed graph (that's the way Anton Kotzig originally described his game) showing the successive places which may be occupied when the move set is $\{1, 2\}$ and you play on a 7-place strip. What happens? By symmetry we can assume that the first player plays on 0, and then he can win as follows:

2nd	1st	2nd	1st	2nd	1st
1?	3!	~	6!		
2?	4!	5?	6	1	3.
		6?	1	3	5.

The sign $\sim$ means "any legal move". Where there's a choice of moves, we've put ! or ? to indicate winning or losing; other moves are forced.

If the move set contains only one move, m, the game is just She-Loves-Me, She-Loves-Me-Not. For if there are n places on the strip, there will be just n/d moves made, where d is the g.c.d. of m and n. If n/d is even, the second player wins; if odd, the first.

If the move set is $\{1, 2\}$, all values of n are $\mathcal{P}$-positions, except for $n = 1, 3$ and 7. We've already seen that $n = 7$ is an $\mathcal{N}$-position, and it's easy to check that $n = 1$ and 3 are, too. Here's a strategy for the second player in all other cases:

	1st	2nd	1st	2nd	1st	2nd		2nd	1st	2nd	1st	2nd
$n = 3k + 2(k \geq 0)$	0?	1!	~	4!	~	7!	$\sim \ldots \sim$	$(3k+1)!$				
$n = 3k(k \geq 2)$	0?	2!	3?	4!	~	7!	$\sim \ldots \sim$	$(3k-2)!$	$3k-1$	1		
			4?	5!	~	8!	$\sim \ldots \sim$	$(3k-1)!$	1	3		
$n = 3k + 1$ ($k = 1, k \geq 3$)	0?	2!	3?	5!	~	8!	$\sim \ldots \sim$	$(3k-1)!$	$3k$	1		
			4?	6!	7?	8!	$\sim \ldots \sim$	$(3k-1)!$	$3k$	1	3	5
					8?	9!	$\sim \ldots \sim$	$(3k)!$	1	3	5	7

(in the last case, if $k = 1$, the move 4 is illegal, so play goes 0? 2! 3 1).

We've got quite used to games which behave regularly after a while, with a few exceptions near the beginning. If the move set is $\{1, 3\}$ the game is *exactly* periodic with period 6. The $\mathcal{N}$-positions are just those with $n \neq 1$ or 3, mod 6. Here's Richard Nowakowski's explanation.

If n is even the first player always plays on even places, so the second player wins, since the move $m = 1$ is always available to him.

If n is odd, then on the first tour of the strip, if a player A responds to a coin placed on p, say, with a coin on $p + 1$, then the other player wins by putting a coin on $p + 2$ (so long as $p + 2 < n$) and A will find himself blocked the next time round. So each player uses the move $n = 3$ as long as he can.

So if $n \equiv 3$, mod 6, the first player will arrive on $n - 3$, forcing the second player to $n - 2$, leaving $n - 1$ to the first player, who wins next time round.

If $n \equiv 1$, mod 6, the first player arrives on $n - 1$. The second time round both players are forced to play on places $p \equiv 2$, mod 3, and the last time round on places $p \equiv 1$, mod 3. Since n is odd, the first player wins.

If $n \equiv 5$, mod 6, the second player wins since first time round he arrives at $n - 2$. The first player now plays on $n - 1$ or on 1, and the corresponding winning replies are 2 and 4.

For the move set $\{2, 3\}$ we leave the reader to verify that

$$n \equiv 0, 1 \text{ and } 4, \text{ mod } 5 \text{ are } \mathcal{P}\text{-positions, } \textit{except} \text{ for } n = 1, 5 \text{ and } 11, \text{ and}$$
$$n \equiv 2 \text{ and } 3, \text{ mod } 5 \text{ are } \mathcal{N}\text{-positions, } \textit{except} \text{ for } n = 2.$$

Omar will also confirm that if the move set is $\{1, 2, 3\}$, then

$$\text{the } \mathcal{P}\text{-positions are } n \equiv 0, 1, 2, \text{ mod } 4, \text{ except for } n = 1 \text{ and } 5, \text{ and}$$
$$\text{the } \mathcal{N}\text{-positions are } n \equiv 3, \text{ mod } 4, \text{ except for } n = 7,$$

and will go on to examine more complicated move sets.

The extension of Kotzig's Nim (or Modular Nim) to more general graphs has been named **Geography** by Fraenkel. Nowakowski & Poole and Hogan & Horrocks have looked at the game which is the product of two cycles. For $3 \times n$ the outcomes are periodic with period 42, and for $4 \times n$ the position is $\mathcal{P}$ just if $n \equiv 11$ mod 12.

Fibonacci Nim

Suppose you play with just one heap of chips, and let the first player take away any number he likes, but not the whole heap. After that each player may take at most twice as many as the previous player took. Who wins?

The $\mathcal{P}$-positions turn out to be heaps with a Fibonacci number

$$u_1 = u_2 = 1, \quad u_3 = 2, \quad u_4 = 3, \quad u_5 = 5, \quad 8, \quad 13, \quad 21, \quad 24, \quad 55, \quad 89, \quad \ldots$$

of chips. Zeckendorf has a remarkable theorem which says that any whole number has a *unique* expression as the sum of *non-neighboring* Fibonacci numbers, for instance

$$54 = 34 + 13 + 5 + 2.$$

If the heap has a *non*-Fibonacci number of chips, the next player can win by taking any number of small terms from such an expansion, provided their total is *less than half* the next largest term. E.g. from a heap of 54 take 2, but not $2 + 5 = 7$ in case your opponent then takes 13.

More Generally Bounded Nim

Suppose the rules are changed very slightly to read

"may take less than twice as many as"

instead of   "may take at most twice as many as;"

does it make much difference? Curiously enough we get the same result as if we had changed the rules to read   "may take no more than."

If the number of chips is a power of 2 it's a $\mathcal{P}$-position in either case; otherwise the next player can win by taking the highest power of 2 which divides the number of chips.

Of course, if the rules say

"may take less than",

then, provided there's more than one chip, you win immediately by taking one, since your opponent then has no legal move. It's a disguise for She-Loves-Me-Constantly.

There are two whole series of such games, in which the rules read

"may take less than k times as many as"

or

"may take at most k times as many as."

For the "less than" games the sequence of $\mathcal{P}$-positions $\{a_n\}$ satisfies the recurrence relation

$$a_{n+1} = a_n + a_{n-l} \text{ for } n \geq n_l,$$

and for the "at most" games the relation is

$$a_{n+1} = a_n + a_{n-m} \text{ for } n \geq n_m,$$

where l, m, n_l and n_m are given in the table:

$k =$	1	2	3	4	5	6	7	...
$l =$	–	0	2	5	7	10	13	...
$m =$	0	1	3	5	7	10	13	...
$n_l =$	–	2	5	13	14	23	28	...
$n_m =$	2	3	6	9	11	19	24	...

To be consistent with the usual labelling of the Fibonacci numbers we start each sequence with $a_2 = 1$. The "less than" sequences continue with

$$a_i = i - 1 \ (2 \leq i \leq k+1), \quad a_i = 2i - k - 2 \ (k+1 \leq i \leq (3k+2)/2), \ldots$$

and the "at most" ones with

$$a_i = i - 1 \ (2 \leq i \leq k+2), \quad a_i = 2i - k - 3 \ (k+2 \leq i \leq (3k+5)/2), \ \ldots$$

but as k increases, it takes longer and longer before the sequences settle down. Omar, having stayed with us so far, will doubtless find the exact way in which these sequences and the above table continue.

Epstein's Put-or-Take-a-Square Game

This is also played with just one heap of chips. At each turn there are just two options: to add or take away the largest perfect square number of chips that there is in the heap. For example, if the number in the heap is a perfect square other than 0, the next player can win by taking the whole heap.

This is a loopy game! If we start from a heap of 2, the legal moves are to add or subtract 1. The first player won't take 1, leaving a perfect square, so he adds 1 to make 3. For the same reason his opponent doesn't *add* 1, so he takes 1 and the game is drawn.

But 5 is a $\mathcal{P}$-position since 5 ± 4 are both squares! And $4 \times 5 = 20$, $9 \times 5 = 45$, $16 \times 5 = 80$ are also $\mathcal{P}$-positions; why not 125? A slightly more interesting $\mathcal{P}$-position is 29. The next player won't *subtract* 25, but when he adds it to make 54, his opponent can go to 5 and win.

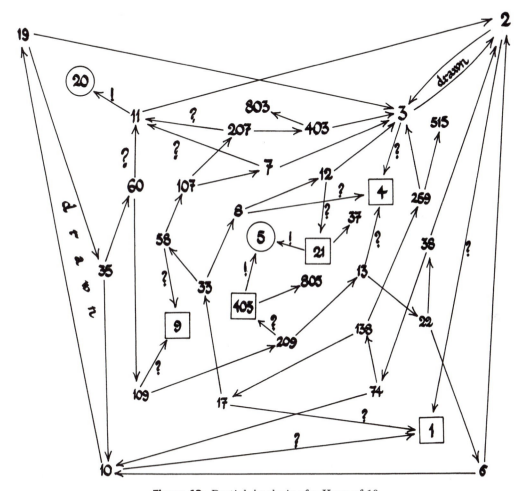

Figure 13. Partial Analysis of a Heap of 10.

Figure 13 shows part of the analysis of games starting from 10. If you continue the figure you'll soon realize why we don't give a complete analysis of Epstein's game. Squares and other $\mathcal{N}$-positions are in square boxes, $\mathcal{P}$-positions are circled.

Here is a list which includes all $\mathcal{P}$-positions of remoteness ≤ 14 below 5000, and a few of the more interesting $\mathcal{N}$-positions, our original list having been augmented by Thea van Roode.

$\mathcal{P}$-positions	$\mathcal{N}$-positions
Remoteness 0 : 0	
	Remoteness 1 : all squares
Remoteness 2 : 5, 20, 45, 80, 145, 580, 949, 1305, 1649, 2320, 3625, 4901, 5220, ...	
	Remoteness 3 : 11, 14, 21, 30, 41, 44, 54, 69, 86, 105, 120, 126, 141, 149, 164, 174, 189, 201, 216, 230, 261, 291, 294, 329, 366, ...
Remoteness 4 : 29, 101, 116, 135, 165, 236, 404, 445, 540, 565, 585, 845, 885, 909, 944, 954, 975, 1125, 1310, 1350, 1380, 1445, 1616, 1654, 1669, 2325, 2340, 2405, 2541, 2586, 2705, 3079, 3150, 3185, 3365, 3380, 3405, 3601, 3630, 3705, 4239, 4921, 4981, 5225, 5265, ...	
	Remoteness 5 : 52, 71, 84, 208, 254, 284, 296, 444, ...
Remoteness 6 : 257, 397, 629, 836, 1177, 1440, 1818, 1833, 1901, 1937, 1988, 2210, 2263, 2280, 2501, 2516, 2612, 2845, 2861, 3039, 3188, 3389, 3621, 3654, 3860, 4053, 4105, 4541, 4693, 4708, 4813, 4930, ...	
	Remoteness 7 : 136, 436, 601, 918, 1291, ...
Remoteness 8 : 477, 666, 5036,...	*Remoteness* 9 : 252, 342,...
Remoteness 10 : 173,...	
	Remoteness 11 : 92,...
Remoteness 12 : 3341, 3573, 3898, 4177, 4229, 4581,...	
	Remoteness 13 : 1809, 1962,...
Remoteness 14 : 1918,...	

If you want to know how to win from a heap of 92, look in the Extras.

The misère form of the game is uninteresting, because the increasing move is always available.

Tribulations and Fibulations

What happens if we use another system of numbers instead of squares? An easy case is $2^k - 1$, but more interesting ones have been suggested to us, namely the triangular numbers 1, 3, 6, 10, 15, 21, ... (Simon Norton) and the Fibonacci numbers plus one, 1, 2, 3, 4, 6, 9, 14, 22, 35, ... (Mike Guy). See the Extras.

Third One Lucky

Ordinary Nim ends when a player takes the last stick. Misère Nim may be thought of as over when only one stick remains. What happens if we say the game's over when exactly two sticks remain, the winner being the player who takes the third last stick? Even the three-heap version of this game is quite hard.

If there are m sticks in the first heap, and n in the second, there is a unique size for the third heap to make a $\mathcal{P}$-position. For fixed m, this size is eventually arithmetico-periodic in n. The periods for

$$m = 1\ 2\ 3\ 4\ \ 5\ \ 6\ \ 7\ 8\ 9\ 10\ 11\ 12\ 13\ 14\ 15\ 16\ 17\ \ 18\ 19 \ldots$$
$$\text{are} \ \ \ \ 1\ 2\ 4\ 2\ 12\ 12\ 12\ 8\ 8\ 10\ 60\ 60\ 84\ 84\ 84\ 16\ 18\ 180\ 20 \ldots$$

Hickory, Dickory, Dock

Dean Hickerson suggested this game in which a move replaces a heap of n by three heaps of sizes

$$k,\ n-k,\ n-2k \ \text{where} \ 1 \le k \le \frac{1}{2}n.$$

It looks rather like Turnips in the previous chapter, but in fact the nim-values for $n = 1, 2, 3, \ldots$ are the exponents, $0, 1, 0, 2, 0, 1, 0, 3, 0, \ldots$ of the nim-values (Fig. 7 of Chapter 14) for the Ruler Game.

D.U.D.E.N.E.Y

D.U.D.E.N.E.Y is a game,

Deductions Unfailing, Disallowing Echoes, Not Exceeding Y;

a particular case of which was described by Dudeney as "The 37 Puzzle Game".

From a single heap either player may subtract a number from 1 to Y:

"Not Exceeding Y"

except that the immediately previous deduction may not be repeated:

"Disallowing Echoes"

and you win if you can always move:

"Deductions Unfailing",

but at some stage your opponent cannot.

If echoes were *not* disallowed, the $\mathcal{P}$-positions would be the multiples of Y+1 and the winner would always follow a deduction of X by one of Y+1−X. This strategy still works for D.U.D.E.N.E.Y when Y is even because it is impossible for Y+1−X to equal X. So we'll suppose from now on that Y is odd.

Here are the good moves from N when Y= 3:

from N =	0	1	2	3	4	5	6	7	8	9	10	11	...
deduct	?	1	1,2	3	?	1	1,2,3	3	?	1	1,2,3	3	...

and here they are for $Y = 5$:

from N =	0	1	2	3	4	5	6	7	8	9	10	11	12	
deduct	?	1	1,2	3	4	5	**3**	?	1,4	2,3	3,5	4	5	
from N =	13	14	15	16	17	18	19	20	21	22	23	24	25	...
deduct	?	1	1,2,4	3	4,5	5	**3**	?	1	1,2,3	3,5	4	5	...

For $Y=3$ there is an ultimate period of 4; for $Y=5$ a period of 13. The easiest way to win is to move to one of those **pearls** among numbers, which have a ? entry, indicating that the next player has no good move at all. Pearls are $\mathcal{P}$-positions no matter what the previous move, but there are other $\mathcal{P}$-positions in which the only winning move is disallowed by the echo rule.

In general the pearls are spaced at intervals of either

$$E \;=\; Y + 1, \text{ the next even number after Y, or}$$
$$D \;=\; Y + 2, \text{ the next odd one.}$$

For, if P is a pearl then after most moves from $P + E$ or $P + D$ we can go immediately to P. The only exceptions are the moves from

$$P + E \text{ to } P + \tfrac{1}{2}E \ \text{ and } P + D \text{ to } P + E.$$

If the first of these is a bad move, $P + E$ is a pearl, and if it's a good one, $P + D$ is a pearl.

From a given position there's usually only one move which will prevent your opponent from reducing to an earlier pearl, though sometimes there can be two. So it's fairly easy to determine the status of the critical move

$$P + E \text{ to } P + \tfrac{1}{2}E$$

by searching back along one or two alleys. In the case $Y=5$, the critical moves from

$$13 \text{ to } 10 \text{ and } 26 \text{ to } 23$$

are bad, because they can be answered by

$$10 \text{ to } 5 \text{ and } 23 \text{ to } 18$$

but those from

$$6 \text{ to } 3 \text{ and } 19 \text{ to } 16$$

are good, and are indicated by bold **3**'s.

Strings of Pearls

Knowing the pearls helps you to win the game: move directly to a pearl if you can, and otherwise prevent your opponent from doing so. So all we need tell you is the sequence of D's and E's separating the pearls. These are (ultimately—see entries 55 and 95) periodic. In Table 6 the periods are in parentheses, and $r \geq 0$. Much of the work was done by John Selfridge and Roger Eggleton.

In his chapter on subtraction games, Schuh discusses this game, together with its misère form, and the two variants in which the outcome is changed if play terminates at 1.

Y	Pearl-string	Y	Pearl-string
3 or $8r+3$	(E)EEE ...	41	(DDDEDE) ...
5 or $8r+5$	(DE)DEDE ...	55	DD(EDE)EDE ...
7	(DEE)DEE ...	63 or $128r+63$	(E)EE
9	(DDE)DDE ...	65 or $128r+65$	(DDDE)DDDE ...
15 or $32r+15$	(E)EE ...	71 or $64r+71$	(DE)DEDE ...
17 or $32r+17$	(DDE)DDE ...	73 or $128r+73$	(DDEDE)DDEDE ...
23	(DDEDDDEE) ...	87 or $128r+87$	(DDE)DDE ...
25 or $32r+25$	(DE)DE ...	95	DDEE(DDE)DDE ...
31 or $128r+31$	(DEE)DEE ...	97	(DDEDDE) ...
33	(DDDEDDE) ...	103	(DE)DEDE ...
39 or $128r+39$	(DEE)DEE ...	105 or $128r+105$	(DE)DEDE ...

Table 6. Strings of Pearls for D.U.D.E.N.E.Y for $\frac{53}{64}$ of the odd Values of Y.

Schuhstrings

Prof. Schuh also discusses the variation in which 0 is a permissible deduction, but the person who first gets to 0 wins.

Starting from any positive number n you'll always have at least one good move, because if no positive deduction wins for you, you can deduct 0 and present your opponent with a similar situation, except that 0 is now illegal. How can a positive deduction

$$n \text{ to } n-g$$

possibly be a good move? The only move it prohibits from $n-g$ is

$$n-g \text{ to } n-2g$$

and so this must be the unique good move from $n-g$. But then similarly

$$n-2g \text{ to } n-3g,$$
$$n-3g \text{ to } n-4g,$$
$$\dots\dots\dots\dots\dots\dots\dots\dots\dots\dots\dots$$

must be good moves, the last of which must be from

$$g \text{ to } 0.$$

A positive deduction g can therefore only be good at a **string**,

$$g, 2g, 3g, \dots, kg, \dots,$$

of multiples of g. It will be good from $(k+1)g$ if and only if it was the *unique* good move from kg. The first multiple of g from which there's another good move will terminate the g-string. Thus, when the permissible deductions are 0, 1, 2, 3, 4, 5, we find that the good moves are:

from n =	1	2	3	4	5	6	7	8	9	10	11	12	13	14	15	16	17	18	19	20	...
deduct	1	1,2]	3	4	5	3	0	4	3	5	0	3,4]	0	0	5	0	0	0	0	5	...

The 1- and 2-strings terminate at 2 and the 3- and 4-strings at 12, but the 5-string continues indefinitely.

In general, if there are two or more numbers

$$a, b, \ldots$$

whose strings have not yet terminated, then the first number to occur in two or more strings will terminate those strings. At most one string continues forever. In Table 7 an entry (a, b) means that the a- and b-strings terminate at their l.c.m., while an entry $g\infty$ corresponds to an infinite g-string, and is only relevant when the largest deduction is odd. It is not known whether there is any Schuhstring game in which three or more strings terminate simultaneously.

2 or 3	4 or 5	6 or 7	8 or 9	10 or 11	12 or 13	14 or 15
(1,2)	(1,2)	(1,2)	(1,2)	(1,2)	(1,2)	(1,2)
3∞	(3,4)	(3,6)	(3,6)	(3,6)	(3,6)	(3,6)
	5∞	(4,5)	(4,8)	(4,8)	(4,8)	(4,8)
(1,2)		7∞	(5,7)	(5,10)	(5,10)	(5,10)
(3,6)		(1,2)	9∞	(7,9)	(9,12)	(7,14)
(4,8)		(3,6)	(1,2)	11∞	(7,11)	(9,12)
(5,10)		(4,8)	(3,6)	(1,2)	13∞	(11,13)
(7,14)		(5,10)	(4,8)	(3,6)	(1,2)	15∞
(9,18)		(7,14)	(5,10)	(4,8)	(3,6)	(1,2)
(11,22)		(9,18)	(7,14)	(5,10)	(4,8)	(3,6)
(12,24)		(11,22)	(9,18)	(7,14)	(5,10)	(4,8)
(13,26)		(12,24)	(11,22)	(9,18)	(7,14)	(5,10)
(15,20)		(15,20)	(12,16)	(12,16)	(9,18)	(7,14)
(21,27)	(16,17)	(13,16)	(15,20)	(15,20)	(12,16)	(9,12)
(16,17)	(19,21)	(17,19)	(13,17)	(11,13)	(11,13)	(11,13)
(19,23)	(23,25)	(21,23)	(19,21)	(17,19)	(15,17)	(15,16)
25∞		25∞	23∞	21∞	19∞	17∞
27	26	25 or 24	23 or 22	21 or 20	19 or 18	17 or 16

Table 7. Schuhstrings Corresponding to Various Maximum Deductions.

The Princess and the Roses

When we originally planned *Winning Ways*, the Princess was to have had a chapter all to herself, but as with other beautiful and intriguing women, we probably dallied too long in her company and it now seems more discreet to limit our memoirs to a brief résumé of our rencontres. Perhaps our narrative will steer a course between the original bare account of Prof. Schuh and the later flights of fancy of Monsieur Filet de Carteblanche.

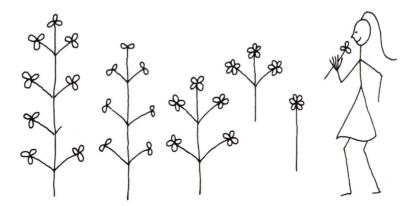

Figure 14. Princess Romantica Smells Charming Charles's Rose.

The Princess Romantica is known to have had two princely suitors, Handsome Hans and Charming Charles. Each suitor went in turn to the rose-garden and would bring back a rose, or two roses from different bushes. In Fig. 14 you can see the Princess smelling the beautiful rose that Charming Charles has just brought from the largest bush. Eventually one suitor, finding himself unable to bring her a rose because none was left in the garden, crept despondently away, and left the other to claim her hand as in Fig. 15. Which was the lucky Prince?

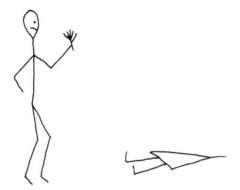

Figure 15. Who Won the Hand of the Princess?

Of course, you can play the game as a heap game in which the legal move is to take any one chip, or any two, one from each of two distinct heaps. Prof. Schuh showed that a worldly prince in a 5-bush garden should always arrange that when the numbers are put in descending order, they form one of the patterns

even-even-even-even-even,
even-odd-odd-odd-odd,
odd-even-even-odd-odd,
odd-odd-odd-even-even.

Obviously Charles knew what he was doing and Prof. Schuh's researches leave little doubt that he must be the man depicted in Fig. 15.

You can see from Prof. Schuh's rule that when there's only a small number of bushes (which may contain a large number of roses).

<table><tr><td>

parity considerations are paramount.

</td></tr></table>

However, when there are many few-rose bushes then

<table><tr><td>

it is triality that triumphs

</td></tr></table>

because the $\mathcal{P}$-positions ultimately are just those in which the total number of roses is a multiple of three.

This reveals itself by the final subscript 3's in Tables 8, 9 and 10 which respectively list all $\mathcal{P}$-positions of the forms

$$3^x 2^y 1^z \qquad \text{or} \qquad a.2^y 1^z \qquad \text{or} \qquad a.b.1^z,$$

i.e.

$$\left.\begin{array}{l} x \text{ 3-rose bushes} \\ y \text{ 2-rose bushes} \\ z \text{ 1-rose bushes} \end{array}\right\} \text{or} \left.\begin{array}{l} 1 \ a\text{-rose bushes} \\ y \text{ 2-rose bushes} \\ z \text{ 1-rose bushes} \end{array}\right\} \text{or} \begin{array}{l} 1 \ a\text{-rose bush} \\ 1 \ b\text{-rose bush} \\ z \text{ 1-rose bushes} \end{array}$$

In these tables an entry n_3 represents all numbers of the infinite arithmetic progression

$$n, n+3, n+6, n+9, \ldots$$

while $m_d n$ represents the finite progression

$$m, m+d, m+2d, \ldots, n,$$

and so on; for example, the entry $_6 7_5 17_3$ represents 1, 7, 12, 17, 20, 23, 26, 29, . . .

x \ y	0	1	2	3	4	5	6	7	8	9	10	11
0	0_3	$_44_3$	$_55_3$	0_3	$_44_3$	$_55_3$	0_3	$_44_3$	$_55_3$	0_3	$_44_3$	$_55_3$
1	$_46_3$	4_3	2_3	3_3	4_3	2_3	3_3	4_3	2_3	3_3	4_3	2_3
2	$_56_3$	4_3	2_3	3_3	1_3	2_3	3_3	1_3	2_3	3_3	1_3	2_3
3	0_3	$_44_3$	$_55_3$	0_3	1_3	2_3	0_3	1_3	2_3	0_3	1_3	2_3
4	3_3	4_3	2_3	0_3	1_3	2_3	0_3	1_3	2_3	0_3	1_3	2_3
5	$_56_3$	4_3	2_3	0_3	1_3	2_3	0_3	1_3	2_3	0_3	1_3	2_3
6	0_3	1_3	2_3	0_3	1_3	2_3	0_3	1_3	2_3	0_3	1_3	2_3
7	0_3	1_3	2_3	0_3	1_3	2_3	0_3	1_3	2_3	0_3	1_3	2_3

Table 8. $\mathcal{P}$-positions of Type $3^x 2^y 1^z$. Entries Are Sets of Values of z.

y \ a	0	1	2	3	4	5	6	7	8	9	10	11
0	0_3	2_3	$_44_3$	$_46_3$	$_48_3$	$_410_3$	$_412_3$	$_414_3$	$_416_3$	$_418_3$	$_420_3$	$_422_3$
1	$_44_3$	3_3	$_55_3$	4_3	$_66_3$	4_48_3	$_66_410_3$	4_412_3	$_66_414_3$	4_416_3	$_66_418_3$	4_420_3
2	$_55_3$	4_3	0_3	2_3	$_44_3$	$_46_3$	$_48_3$	$_410_3$	$_412_3$	$_414_3$	$_416_3$	$_418_3$
3	0_3	2_3	$_44_3$	3_3	$_55_3$	4_3	$_66_3$	4_48_3	$_66_410_3$	4_412_3	$_66_414_3$	4_416_3
4	$_44_3$	3_3	$_55_3$	4_3	0_3	2_3	$_44_3$	$_46_3$	$_48_3$	$_410_3$	$_412_3$	$_414_3$
5	$_55_3$	4_3	0_3	2_3	$_44_3$	3_3	$_55_3$	4_3	$_66_3$	4_48_3	$_66_410_3$	4_412_3
6	0_3	2_3	$_44_3$	3_3	$_55_3$	4_3	0_3	2_3	$_44_3$	$_46_3$	$_48_3$	$_410_3$
7	$_44_3$	3_3	$_55_3$	4_3	0_3	2_3	$_44_3$	3_3	$_55_3$	4_3	$_66_3$	4_48_3

Table 9. $\mathcal{P}$-positions of Type $a.2^y 1^z$. Entries Are Sets of Values of z.

b \ a	0	1	2	3	4	5	6	7	8	9	10	11
0	0_3	2_3	${}_4 4_3$	${}_4 6_3$	${}_4 8_3$	${}_4 10_3$	${}_4 12_3$	${}_4 14_3$	${}_4 16_3$	${}_4 18_3$	${}_4 20_3$	${}_4 22_3$
1	2_3	1_3	3_3	${}_4 5_3$	${}_4 7_3$	${}_4 9_3$	${}_4 11_3$	${}_4 13_3$	${}_4 15_3$	${}_4 17_3$	${}_4 19_3$	${}_4 21_3$
2	${}_4 4_3$	3_3	${}_5 5_3$	4_3	6_3	${}_4 8_3$	${}_6 10_3$	${}_4 12_3$	${}_6 14_3$	${}_4 16_3$	${}_6 18_3$	${}_4 20_3$
3	${}_4 6_3$	${}_4 5_3$	4_3	${}_5 6_3$	${}_5 8_3$	${}_6 7_3$	${}_6 9_3$	${}_6 7{}_4 11_3$	${}_6 9{}_4 13_3$	${}_6 7{}_4 15_3$	${}_6 9{}_4 17_3$	${}_6 7{}_4 19_3$
4	${}_4 8_3$	${}_4 7_3$	6_3	${}_5 8_3$	${}_5 10_3$	${}_5 9_3$	${}_6 6{}_5 11_3$	${}_6 10_3$	${}_6 12_3$	${}_6 10{}_4 14_3$	${}_6 12{}_4 16_3$	${}_6 10{}_4 18_3$
5	${}_4 10_3$	${}_4 9_3$	${}_4 8_3$	${}_6 7_3$	${}_5 9_3$	${}_5 11_3$	${}_5 13_3$	${}_6 7{}_5 12_3$	${}_6 9{}_5 14_3$	${}_6 13_3$	${}_6 15_3$	${}_6 13{}_4 17_3$
6	${}_4 12_3$	${}_4 11_3$	${}_6 10_3$	${}_6 9_3$	${}_6 6{}_5 11_3$	${}_5 13_3$	${}_5 15_3$	${}_5 14_3$	${}_6 6{}_5 16_3$	${}_6 10{}_5 15_3$	${}_6 12{}_5 17_3$	${}_6 16_3$
7	${}_4 14_3$	${}_4 13_3$	${}_4 12_3$	${}_6 7{}_4 11_3$	${}_6 10_3$	${}_6 7{}_5 12_3$	${}_5 14_3$	${}_5 16_3$	$3{}_5 18_3$	${}_6 7{}_5 17_3$	${}_6 9{}_5 19_3$	${}_6 13{}_5 18_3$
8.	${}_4 16_3$	${}_4 15_3$	${}_6 14_3$	${}_6 9{}_4 13_3$	${}_6 12_3$	${}_6 9{}_5 14_3$	${}_6 6{}_5 16_3$	$3{}_5 18_3$	${}_5 20_3$	${}_5 19_3$	${}_6{}_5 21_3$	${}_6 10{}_5 20_3$
9	${}_4 18_3$	${}_4 17_3$	${}_4 16_3$	${}_6 7{}_4 15_3$	${}_6 10{}_4 14_3$	${}_6 13_3$	${}_6 10{}_5 15_3$	${}_6 7{}_5 17_3$	${}_5 19_3$	${}_5 21_3$	${}_5 23_3$	${}_6 7{}_5 22_3$
10	${}_4 20_3$	${}_4 19_3$	${}_6 18_3$	${}_6 9{}_4 17_3$	${}_6 12{}_4 16_3$	${}_6 15_3$	${}_6 12{}_5 17_3$	${}_6 9{}_5 19_3$	${}_6{}_5 21_3$	${}_5 23_3$	${}_5 25_3$	${}_5 24_3$
11	${}_4 22_3$	${}_4 21_3$	${}_4 20_3$	${}_6 7{}_4 19_3$	${}_6 10{}_4 18_3$	${}_6 13{}_4 17_3$	${}_6 16_3$	${}_6 13{}_5 18_3$	${}_6 10{}_5 20_3$	${}_6 7{}_5 22_3$	${}_5 24_3$	${}_5 26_3$

Table 10. $\mathcal{P}$-positions of Type $a.b.1^z$. Entries Are Sets of Values of z.

The tables also illustrate that there are places between the parity and triality regions in which the outcome depends on considerations mod 4 and 5, so that

> quaternity's a quality,

> quinticity can be quintessential ,

and there are even hints that

> sex may be significant;

but we have explored other regions in which it sadly seems that

> randomness reigns.

In his first paper on this subject M. de Carteblanche asks for a code of behavior for princes in a 6-bush garden wherein there's a bush with only 1 rose. You'll find one in the Extras. In a second paper he further describes how the princes, after their weddings to Romantica and her even more beautiful younger sister Belladonna, transformed the rose game into a different one with chocolates and discovered some more interesting games to play.

One-Step, Two-Step

This is the strip game in which arbitrarily many coins are allowed on a square, and the legal move is to make either one step or two steps, a **step** being to move a single coin just one space leftwards. The coins moved in the two steps of a 2-step move may be the same or different.

Letting a_n be the number of coins on square n, we can ask when

$$a_0 a_1 a_2 \ldots$$

represents a $\mathcal{P}$-position. The answer certainly won't depend on a_0 since coins on square 0 will never be moved again.

There's a surprising connexion between this game and our previous one. In fact the position described above behaves exactly like

$$a_1 + a_2 + a_3 + \ldots + a_n, \quad a_2 + a_3 + \ldots + a_n, \quad a_3 + \ldots + a_n, \ldots, a_{n-1} + a_n, \quad a_n$$

in the Princess-and-Roses game! We leave it to Omar to work out why.

So when your coins are all in the first 6 places, you can translate Prof. Schuh's rules to give the $\mathcal{P}$-positions, which are

$$?eeeee, \quad ?deeed, \quad ?deded, \quad ?eedee,$$

where e means even, d means odd and $?$ means anything.

More on Subtraction Games

Since $\mathcal{G}(n)$ for a heap of n beans in the subtraction game (see Chapter 4, Vol 1.)

$$S(s_1, s_2, \ldots, s_k)$$

depends only on k earlier values, namely

$$\mathcal{G}(n - s_1), \mathcal{G}(n - s_2), \ldots, \mathcal{G}(n - s_k),$$

we see that $\mathcal{G}(n) \leq k$. Moreover this sequence of k values must eventually repeat so the nim-sequences of all subtraction games are (ultimately) periodic. But the bound on the length of the period given by this argument seems astronomical when compared with the facts. Can you find something nearer the truth?

We have seen that if the g.c.d. $(s_1, s_2, \ldots, s_k)$ is $d > 1$, then the game is just the d-plicate of a simpler game. Thus $S(s_1)$ is the s_1-plicate of $S(1)$, She-Loves-Me, She-Loves-Me-Not, and so has period $2s_1$ and nim-sequence $\dot{0}.00 \ldots 0111 \ldots \dot{1}$.

We can also analyze $S(s_1, s_2)$ and $S(s_1, s_2, s_1 + s_2)$ completely. Write

$$s_1 = a, \quad s_2 = b = 2ha \pm r \quad \text{for } 0 \leq r \leq a$$

and suitable h. After the g.c.d. remark we needn't consider $r = 0$ or a unless $a = 1$.

$S(1, 2h)$ has period $2h + 1$ and nim-sequence $\dot{0}.10101 \ldots 01\dot{2}$, and $S(1, 2h + 1) = S(1)$. (In fact $s_1 = 1$ and all s_i odd gives She-Loves-Me, She-Loves-Me-Not.)

For $a > 1$ the period of $S(a, b)$ contains $a + b$ digits, alternating blocks of a 0's and a 1's, except that the last $a - r$ 0's are replaced by 2's, where r is as above. For example: $a = 3$, $r = 1$; the nim-sequences for $S(3, 11)$ and $S(3, 13)$ are

$$\dot{0}.0011100011\dot{2} \quad \text{and} \quad \dot{0}.001110001110\dot{2}\dot{1}.$$

Here is a general method for analyzing $S(s_1, s_2 \ldots, s_k)$. Write the numbers in $k+1$ columns. The first row is

$$0, s_1, s_2, \ldots_{.k}.$$

Each later row is of the form

$$l, \ l + s_1, \ l + s_2, \ \ldots \ l + s_k,$$

where l is the least whole number which hasn't appeared in earlier rows. The table will eventually become periodic in that a block of c consecutive rows can be obtained from the preceding block of c by adding p to all the entries, for a suitable c and p.

The first column contains all numbers n for which $\mathcal{G}(n) = 0$, and the second, by Ferguson's pairing property, just those for which $\mathcal{G}(n) = 1$. Later columns contain numbers for which $\mathcal{G}(n) \geq 2$, apart from repetitions of entries in the second column. We illustrate with $S(1, b, b + 1)$. If b is even (Fig. 16(a)) there are no such repetitions, the period is $2b$ and the nim-sequence is

$$\dot{0}.101\ldots012323\ldots2\dot{3}.$$

$\mathcal{G}(n) =$	0	1	2	3		$\mathcal{G}(n) =$	0	1	3	2	except that
$n =$	0	1	10	11		$n =$	0	1	9	10	$\mathcal{G}(9) = 1$
	2	3	12	13			2	3	11	12	
	4	5	14	15			4	5	13	14	
	6	7	16	17			6	7	15	16	
	8	9	18	19			8	9	17	18	
	20	21	30	31			19	20	28	29	$\mathcal{G}(28) = 1$
	22	23	32	33			21	22	30	31	
	24	25	34	35			23	24	32	33	
	26	27	36	37			25	26	34	35	
	28	29	38	39			27	28	36	37	
	40	41	50	51			38	39	47	48	$\mathcal{G}(47) = 1$
	42	...					40	...			
		(a)						(b)			

Figure 16. The Subtraction Games S(1,10,11) and S(1,9,10).

If b is odd (Fig. 16b) there is one repetition $(9, 28, 47, \ldots)$ in each period, whose length is $2b + 1$. The nim-sequence is as before, but with the final 3 omitted:

$$\dot{0}.101\ldots012323\ldots\dot{2}.$$

$G(n)=$		0	1	0	1
$n=$		0	a	【$2ha-r$】	$(2h+1)a-r$
		1	$a+1$	【$2ha-r+1$】	$(2h+1)a-r+1$
		2	$a+2$	【$2ha-r+2$】	$(2h+1)a-r+2$
		$\cdots$	$\cdots$	【$\cdots$】	$\cdots$
		$r-1$	$a+r-1$	【$2ha-1$】	$(2h+1)a-1$
		r	$a+r$	$2ha$	$(2h+1)a$
		$\cdots$	$\cdots$	$\cdots$	$\cdots$
		$a-1$	$2a-1$	$(2h+1)a-r-1$	$(2h+2)a-r-1$
		$2a$	$3a$	$(2h+2)a-r$	$(2h+3)a-r$
		$2a+1$	$3a+1$	$(2h+2)a-r+1$	$(2h+3)a-r+1$
		$\cdots$	$\cdots$	$\cdots$	$\cdots$
		$3a-1$	$4a-1$	$(2h+3)a-r-1$	$(2h+4)a-r-1$
		$4a$	$5a$	$(2h+4)a-r$	$(2h+5)a-r$
		$4a+1$	$5a+1$	$(2h+4)a-r+1$	$(2h+5)a-r+1$
		$\cdots$	$\cdots$	$\cdots$	$\cdots$
		$5a-1$	$6a-1$	$(2h+5)a-r-1$	$(2h+6)a-r-1$
		$\cdots$	$\cdots$	$\cdots$	$\cdots$
		$(2h-2)a$	$(2h-1)a$	$(4h-2)a-r$	$(4h-1)a-r$
		$(2h-2)a+1$	$(2h-1)a+1$	$(4h-2)a-r+1$	$(4h-1)a-r+1$
		$\cdots$	$\cdots$	$\cdots$	$\cdots$
		$(2h-1)a-r-1$	$2ha-r-1$	$(4h-1)a-2r-1$	$4ha-2r-1$
		$(2h-1)a-r$	【$2ha-r$】	$(4h-1)a-2r$	$4ha-2r$
		$\cdots$	【$\cdots$】	$\cdots$	$\cdots$
		$(2h-1)a-1$	【$2ha-1$】	$(4h-1)a-r-1$	$4ha-r-1$

Figure 17. Analysis of $S(a,b,a+b)$, $b=2ha-r$, $0<r<a$, $(a,b)=1$.

To complete the analysis of $S(a,b,a+b)$ note that the case $a>1$, $b=2ha-r$, $0<r<a$ is fairly straightforward. It is illustrated by a period of ha rows in Fig. 17; there are r repetitions (boxed in the figure) so the period is $4ha-r=2b+r$. The period comprises h blocks of a 0's and a 1's followed by $h-1$ blocks of a 2's and a 3's, then a 2's and $a-r$ 3's.

The case $b=2ha+r$, $0<r<a$, is more complicated. The period is a times as long, $(2b+r)a$. We illustrate it with the particular case $a=5$, $b=43$, $h=4$, $r=3$ in Fig. 18. The ar ($=15$) repetitions are shown boxed.

In either of the cases $b=2ha\pm r$, the ith value of n for which $\mathcal{G}(n)=0$ is

$$n_i = i + \left\lfloor \frac{i}{a} \right\rfloor a + \left\lfloor \frac{2i}{b+r} \right\rfloor b$$

and the value for which $\mathcal{G}(n)=1$ is $a+n_i$ by Ferguson's pairing property.

0	5	43	48		93	98	136	141						269	274	312	317					
1	6	44	49		94	99	137	142						270	275	313	318					
2	7	45	50		95	100	138	143						271	276	314	319					
3	8	46	51		96	101	139	144						272	277	315	320					
4	9	47	52		97	102	140	145						273	278	316	321					
10	15	53	58		103	108	146	151		186	191	229	234	279	284	322	327		362	367	405	410
11	16	54	59		104	109	147	152		187	192	230	235	280	285	323	328		363	368	406	411
12	17	55	60		105	110	148	153		188	193	231	236	281	286	324	329		364	369	407	412
13	18	56	61		106	111	149	154		189	194	232	237	282	287	325	330		365	370	408	413
14	19	57	62		107	112	150	155		190	195	233	238	283	288	326	331		366	371	409	414
20	25	63	68		113	118	156	161		196	201	239	244	289	294	332	337		372	377	415	420
21	26	64	69		114	119	157	162		197	202	240	245	290	295	333	338		373	378	416	421
22	27	65	70		115	120	158	163		198	203	241	246	291	296	334	339		374	379	417	422
23	28	66	71		116	121	159	164		199	204	242	247	292	297	335	340		375	380	418	42?
24	29	67	72		117	122	160	165		200	205	243	248	293	298	336	341		376	381	419	42<
30	35	73	78		123	128	166	171		206	211	249	254	299	304	342	347		382	387	425	43(
31	36	74	79		124	129	167	172		207	212	250	255	300	305	343	348		383	388	426	43
32	37	75	80		125	130	168	173		208	213	251	256	301	306	344	349		384	389	427	43:
33	38	76	81		126	131	169	174		209	214	252	257	302	307	345	350		385	390	428	43
34	39	77	82		127	132	170	175		210	215	253	258	303	308	346	351		386	391	429	43.
40	45	83	88		133	138	176	181		216	221	259	264	309	314	352	357		392	397	435	44
41	46	84	89							217	222	260	265	310	315	353	358		393	398	436	44
42	47	85	90		177	182	220	225		218	223	261	266						394	399	437	44
					178	183	221	226		219	224	262	267	354	359	397	402		395	400	438	44
86	91	129	134		179	184	222	227						355	360	398	403		396	401	439	44
87	92	130	135		180	185	223	228		263	268	306	311	356	361	399	404					

Figure 18. Analysis of $S(5, 43, 48)$.

Smallest Nim, Largest Nim

In **Smallest Nim** there are two colors of heap, black and grey. A legal move is to take any number of beans from the *smallest* black heap, or from any grey heap. To calculate the nim-value of a position where the smallest black heap contains n beans, and there are k such n-bean heaps, then all black heaps may be thought of as a single grey heap of size n or $n - 1$, according as k is odd or even if these are the only black heaps, but according as k is even or odd if they are any larger black heaps.

In **Largest Nim** the two colors are grey and white and a legal move to take any number of beans from any grey heap or from the *largest* white heap. If any two or more white heaps have the same size, say n, then the value of the position is unchanged if all smaller white heaps are discarded and $n - 1$ beans are removed from each of the remaining heaps of n or more beans. Any even number of 1-heaps may be discarded, and the process repeated, so that an arbitrary position in Largest Nim is equivalent to one in which the white heaps have different sizes. Omar will enjoy calculating the value of two such white heaps, and will conclude that Largest Nim is rather more complicated than Smallest Nim.

Trevor Green, Claudio Baiocchi, and Philippe Fondanaiche found all P-positions for the special cases in which all heaps are black or all heaps are white. Several papers are available on-line at http://www.stetson.edu/efriedma/mathmagic/1100.html at the Math Magic Website, November 2000 .

Moore's Nim$_k$

E. H. Moore suggested the heap game in which the legal move is to reduce the size of any positive number, up to k, of heaps. Thus Nim$_1$ is ordinary Nim, and Nim$_2$ is the game in which you can reduce just *one* or *two* heaps. The theory rather surprisingly involves calculations in base two and in base $k + 1$. You

Expand the numbers in base 2, and

Add them in base $k + 1$, without carrying.

You should move to positions in which this "sum" is zero.

For example, if $k = 2$, and you are confronted with

5 =	101		5 =	101
6 =	110		6 =	110
9 =	1001		3 =	11
10 =	1010		7 =	111
	2222			000

you must reduce the 9 and 10 heaps, replacing them by 3 and 7.

Smith's analysis of Subselective Compounds (Chapter 12) is similar. Nim-values for Moore's game have been found by Jenkyns and Mayberry. Yamasaki has shown that it is tame, and that a position is fickle only if its nonzero heaps are all of size 1, and the number of them is 0 or 1 modulo $k + 1$.

The More the Merrier

Bob Li has suggested that ordinary Nim can be adapted for n players. They take turns in a fixed cyclic order and there are different grades of winner.

First prize goes to the player who makes the last move,
Second prize to the immediately previous player, and so on, until the
Booby prize, which goes to the player who was first unable to move.

Sharing of prizes is not permitted: as soon as the game ends each player must take his prize and set off for his home town without any under-the-table payoffs for help he might have received from some of the other players.

Li was surprised to find that his game was very similar to Moore's. Take the position you're faced with and add the binary expansions of your numbers in base n without carrying. You're the Booby only when the sum is 0.

Moore and More

If we allow each of the n players to reduce any number up to k of the heaps at his move, the theory, also due to Li, is similar. The Booby this time is the player who sees 0 when he adds the binary expansions of the numbers modulo $k(n-1)+1$, without carrying.

We have not discussed games for more than 2 players elsewhere in this book because the stipulations to prevent coalitions are somewhat artificial, and lead to paradoxes of the "surprise exam" type. See the article by Paul Hudson in the References.

There are so many generalizations of Nim with interesting theories that we certainly haven't said the last word on the subject, and so

> This is the way the chapter ends,
> This is the way the chapter ends,
> This is the way the chapter ends,

Not with a Bang but a Whim

Perhaps you might like to play Nim, but on just one occasion one of the two players is allowed, instead of his usual Nim move, to exercise his **Whim** to decide whether the outcome will be decided by normal or by misère play. If we use

> 0-nim to mean normal Nim
> 1-nim to mean misère Nim, and
> 2-nim to mean Whim,

then we can continue the sequence with

> 3-nim = **Trim**,
> 4-nim = **Quam**, etc.

The move in d-nim ("**Denim**"), $d \geq 2$, is

> *either*   to move as in Nim
> *or*   to reduce d (but not both).

This is easy to analyze if you introduce a **quiddity heap**, to keep account of just which game you're playing. Then each of these games becomes like Nim with an extra heap. If $2k \leq d \leq 2^{k+1}$, the quiddity heap behaves like a heap of size d when all other heaps are of size less than 2^{k+1}, but like a heap of size $d-1$ when they're not.

Extras

Did You Win the Silver Dollar?

You did if you moved the coin behind the $ just 2 squares or 1, leaving (3,2,1) or (2, 3,1), depending on which version you're playing. Notice that, in the latter case, there's only one coin between the $ and the bag.

How Was Your Arithmetic?

When you filled in the frieze pattern it should have looked like

```
1    1    1    1    1    1    1    1    1    1    1    1         ...
   1    2    3    2    2    1    4    3    1    2    3         ...
     1    5    5    3    1    3   11    2    1    5    5         ...
   1    2    8    7    1    2    8    7    1    2    8         ...
     1    3   11    2    1    5    5    3    1    3   11         ...
   1    4    3    1    2    3    2    2    1    4    3         ...
     1    1    1    1    1    1    1    1    1    1         1    ...
```

In Put-or-Take-a-Square , 92 Is an $\mathcal{N}$-Position

In Fig. 19 the $\mathcal{P}$-positions are in the centre column, squares on the right and other N-positions on the left.

Tribulations and Fibulations

Norton conjectures that in his game of **Tribulations** no position is drawn and $\mathcal{N}$-positions are more numerous than $\mathcal{P}$-positions in golden ratio. Richard Parker has verified these assertions for numbers < 5000 (for which the calculations sometimes run into the millions). The remoteness ad suspense numbers (which are probably always finite) are shown in Table 11. Play from 51, 2 and 56 is especially interesting; draw diagrams like Fig. 19.

For Mike Guy's game of **Fibulations** we have proved the corresponding assertions and can in fact give a complete analysis. It is well known that any number can be economically expressed as the sum of Fibonacci numbers by the **Zeckendorf algorithm** : always subtract the largest Fibonacci number you can. Less economically we can use the **Secondoff algorithm**: always take off the *second* largest Fibonacci number you can, e.g.

$$100 = 89 + 8 + 3 \text{ (Zeckendorf) or } 55 + 21 + 13 + 5 + 3 + 2 + 1 \text{ (Secondoff)}$$

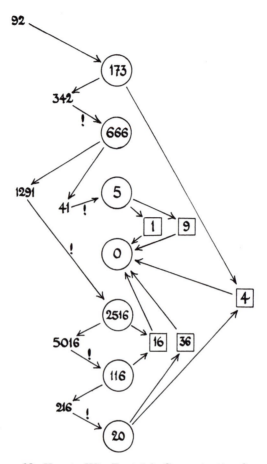

Figure 19. How to Win Epstein's Game starting from 92.

n	1	2	3	4	5	6	7	8	9	10	11	12	13	14	15	16	17	18	19	20	21	22	23	24	25
R	1	2	1	6	3	1	5	3	2	1	2	3	4	3	1	9	3	6	7	8	1	10	3	2	3
S	3	2	1	4	3	1	3	3	2	5	4	5	2	5	1	3	3	2	5	4	3	4	5	2	5

n	26	27	28	29	30	31	32	33	34	35	36	37	38	39	40	41	42	43	44	45	46	47	48	49	50
R	4	5	1	4	3	8	7	5	9	7	1	14	3	4	7	4	2	9	4	1	2	3	4	7	8
S	4	3	5	4	3	2	5	3	5	5	3	4	5	2	5	4	2	5	4	7	2	5	2	5	4

n	51	52	53	54	55	56	57	58	59	60	61	62	63	64	65	66	67	68	69	70	71	72	73	74	75
R	12	16	9	3	1	12	3	14	7	6	4	8	6	3	2	1	6	3	5	7	11	4	13	8	3
S	2	4	5	3	5	4	5	2	5	4	2	4	4	7	2	1	4	3	3	5	5	2	3	4	5

Table 11. Remoteness and Suspense Numbers for Tribulations.

A number is a $\mathcal{P}$-position in Fibulations if and only if

either   its Secondoff expansion ends $3 + 1 + 1$ or $5 + 2 + 1$
or   it is 3 more than a Fibonacci number which is at least 8.

The numbers	0,	11,	5,	8,	13,	21,	34,	55,	89,	144,	...
have remotenesses	0,	8,	2,	2,	2,	4,	6,	8,	10,	12,	...

The remoteness of any other $\mathcal{P}$-position is found by adding twice the number of Secondoff steps you take to get to one of these and the remoteness of an $\mathcal{N}$-position is one more than the smallest remoteness of its $\mathcal{P}$-options. E.g. from 1000 we get to 34 (remoteness 6) after 4 subtractions (of 610, 233, 89 and 34) so 1000 has remoteness $6 + (4 \times 2) = 14$. On the other hand, 1001 has remoteness 3 (move to 13). We believe, and Omar might confirm, that the suspense numbers (which are all finite) and nim-values (which are all 0, 1, 2 or ∞_0) have similar patterns.

Our Code of Behavior for Princes

Our Code of Behavior for Princes in that 6-bush rose-garden with a 1-rose bush is best described by translating it into the One-Step, Two-Step game. It can be checked that all the resulting positions except

?*edede*1	?*eddde*1	?*ddede*1	?*dddde*1
?*ededd*1	?*edddd*1	?*ddedd*1	?*ddddd*1
?0*eeee*1	?0*edee*1	?*ed*0*ed*1	?*dd*0*ee*1

can be moved to one of Schuh's $\mathcal{P}$-positions, and that these 12 classes of position, when joined by possible moves, form the graph of Fig. 20. In the figure a boxed position is $\mathcal{P}$, an unboxed one is $\mathcal{N}$, and

e	means any even number, including 0,
d	means any odd number,
E	means any even number ≥ 2,
D	means any odd number ≥ 3, and
?	means any number,

and the dotted arrow indicates that moves can only be made in that direction. The positions of form ?00*dee*1 and ?00*eee*1 have been omitted from the figure because they cannot be reached from the other ones. To complete the figure, adjoin

?00*dee*1	?00*eee*1
except	except
?00*dEE*1	?00*EEE*1
?001021	?000*E*01
?001041	?002021.

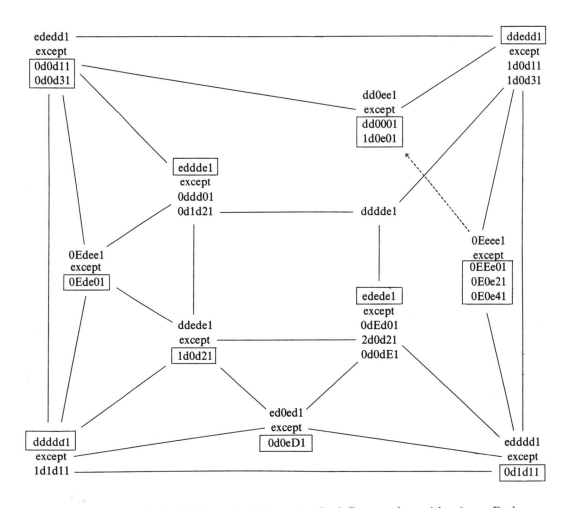

Figure 20. Our Code of Behavior for Princes in 6-bush Rose-gardens with a 1-rose Bush.

References and Further Reading

Richard Austin, Impartial and Partisan Games, M.Sc. thesis, University of Calgary, 1976.

W. W. Rouse Ball and H. S. M. Coxeter, *Mathematical Recreations & Essays*, 12th edn., University of Toronto Press, 1974 (especially pp. 38–39).

E. R. Berlekamp, Unsolved problem #4, in W. T. Tutte (ed.) *Recent Progress in Combinatorics*, Academic Press, New York and London, 1969, pp. 342–343.

E. R. Berlekamp, Some recent results on the combinatorial game called Welter's Nim, *Proc. 6th Conf. Information Sci. and Systems, Princeton*, 1972, 203–204.

F. de Carteblanche, The princess and the roses, *J. Recreational Math.* **3** (1970) 238–239.

F. de Carte Blanche, The roses and the princes, ibid. **7** (1974) 295–298.

J. H. Conway, *On Numbers and Games*, A K Peters, Ltd., Natick, MA, 2001, Chapters 11 and 13, and p. 181.

J. H. Conway and H. S. M. Coxeter, Triangulated polygons and frieze patterns. *Math. Gaz.* **57** (1973) 87–94, 175–183; MR 57 #1254-5.

H. E. Dudeney, *536 Puzzles and Curious Problems*, (ed. Martin Gardner) Chas., N.Y., 1969; #475 The 37 Puzzle Game, 186–187, 392–393.

Robert J. Epp and Thomas S. Ferguson, Remarks on take-away games and Dawson's Game, Abstract 742-90-3, *Notices Amer. Math. Soc.* **24** (1977) A-179.

Jim Flanigan, Generalized two-pile Fibonacci Nim, *Fibonacci Quart*, **16** (1978) 459–469.

A. S. Fraenkel and S. Simonson, Geography, *Theoret. Comput. Sci.* (*Math. Games*), **110** (1993) 197–214; MR 94h:90083.

A. S. Fraenkel, E. R. Scheinerman and D. Ullman, Undirected edge Geography, *Theoret. Comput. Sci.* (*Math. Games*), **112** (1993) 371–381; MR 94a:90043.

A. S. Fraenkel, A. Jaffray, A. Kotzig and G. Sabidussi, Modular Nim, *Theoret. Comput. Sci.* (*Math. Games*), **143** (1995) 319–333; MR 96f:90137.

Richard K. Guy, Anyone for Twopins?, in David Klarner (ed.) *The Mathematical Gardner*, Prindle Weber and Schmidt, 1980.

M. S. Hogan and D. G. Horrocks, Geography played on an n-cycle times a 4-cycle, *Integers: Elec. J. Combin. Number Theory*, **3** (2003) G2 (electron.)

Paul D.C. Hudson, The logic of social conflict: a game-theoretic approach in one lesson, *Bull. Inst. Math. Appl.* **14** (1978) 54–66.

Thomas A. Jenkyns and John P. Mayberry, The skeleton of an impartial game and the nim-function of Moore's Nim$_k$, *Internat. J. Game Theory* **9** (1980) 51–63.

S.-Y. R. Li, N-person Nim and N-person Moore's games, *Internat. J. Game Theory*, **7** (1978) 31–36; MR 58 #4367.

David Moews, Infinitesimals and coin-sliding. in Richard Nowakowski (ed.) *More Games of No Chance*, (Berkeley CA 2000) Math. Sci. Res. Inst. Publ., 42 (2002) Cambridge Univ. Press, Cambridge, UK, 315–327; MR **97h**:90094.

Eliakim H. Moore, A generalization of the game called Nim, *Ann. of Math. Princeton* **2**:11 (1910) 93–94.

J. von Neumann and O. Morgenstern, *Theory of Games and Economic Behavior*, Princeton, 1944.

Richard Nowakowski, ... Welter's Game, Sylver Coinage, Dots-and-Boxes, ..., in Richard K. Guy (ed.) *Combinatorial Games, Proc. Symp. Appl. Math.*, 43(1991), Amer. Math. Soc., Providence, RI, 155–182.

Richard J. Nowakowski and David J. Poole, Geography played on products of cycles, in Richard Nowakowski (ed.) *More Games of No Chance*, (Berkeley CA 2000) *Math. Sci. Res. Inst. Publ.*, 42 (2002) Cambridge Univ. Press, Cambridge, UK, 183–212; MR 97j:90102.

T. H. O'Beirne, *Puzzles and Paradoxes*, Oxford University Press, London, 1965, Chapter 9.

I. C. Pond and D. F. Howells, More on Fibonacci Nim, *Fibonacci Quart.* **3** (1965) 61.

Fred. Schuh, *The Master Book of Mathematical Recreations*, (transl. F. Göbel, from *Wonderlijke Problemen; Leerzaam Tijdverdrijf Door Puzzle en Spel*, W. J. Thieme, Zutphen, 1943; ed. T. H. O'Beirne) Dover, London, 1968. Chapter VI, 131–154; Chapter XII, 263–280.

Allen J. Schwenk, Take-away games, *Fibonacci Quart.* **8** (1970) 225–234, 241; MR 44 #1446.

G. C. Shephard, Additive frieze patterns and multiplication tables, *Math. Gaz.* **60** (1976) 178–184; MR 58 #16353.

Roland Sprague, *Recreations in Mathematics* (trans. T.H. O'Beirne) Blackie, 1963; #14: Pieces to be moved, pp.12–14, 41–42.

R. Sprague, Bemerkungen über eine spezielle Abelsche Gruppe, *Math. Z.* **51** (1947) 82–84; MR 9, 330–331.

C. P. Welter, The advancing operation in a special abelian group, *Nederl. Akad. Wetensch. Proc. Ser. A* 55 = *Indagationes Math.* **14** (1952) 304–314; MR 14, 132.

C. P. Welter, The theory of a class of games on a sequence of squares, in terms of the advancing operation in a special group, ibid. 57 = 16 (1954) 194–200; MR 15, 682; 17, 1436.

Michael J. Whinihan, Fibonacci Nim, *Fibonacci Quart.* **1** (1963) 9–13.

Yôhei Yamasaki, On misère Nim-type games, *J. Math. Soc. Japan*, 32 (1980) 461–475.

Michael Zieve, Take-away games, in Richard Nowakowski (ed.) *Games of No Chance*, (Berkeley CA 1994) Math. Sci. Res. Inst. Publ., 29 (1996) Cambridge Univ. Press, Cambridge, UK, 351–361; MR 97i:90136.

-16-

Dots-and-Boxes

Come, children, let us shut up the box.
William Makepeace Thackeray, *Vanity Fair*, Ch. 67.

I could never make out what those damned dots meant.
Lord Randolph Churchill.

Dots-and-Boxes is a familiar paper and pencil game for two players and has other names in various parts of the world. Two players start from a rectangular array of dots and take turns to join two horizontally or vertically adjacent dots. If a player completes the fourth side of a unit square (**box**) he initials that box and must then draw another line (so that completing a box is a complimenting move). When all the boxes have been completed the game ends and whoever has initialled more boxes is declared the winner.

A player who *can* complete a box is not obliged to do so if he has something else he prefers to do. Play would become significantly simpler were this obligation imposed; see the article by Holladay mentioned in the references.

Figure 1 shows Arthur's and Bertha's first game, in which Arthur started. Nothing was given away in the fairly typical opening until Arthur was forced to make the unlucky thirteenth move, releasing 2 boxes for Bertha. Her last bonus move enabled Arthur to take the bottom 3 boxes, but he then had to surrender the last 4.

This is how most children play, but Bertha is brighter than most. She started the return match with the opening that Arthur had used. He was happy to copy Bertha's replies from that game, and was delighted to see her follow it even as far as that unlucky thirteenth move, which had proved his undoing (Fig. 2). He grabbed those 2 boxes and happily surrendered·the bottom 3, expecting 4 in return. But Bertha astounded him by giving him back 2. He pounced on these, but when he came to make his bonus move, realized he was doubled-crossed!

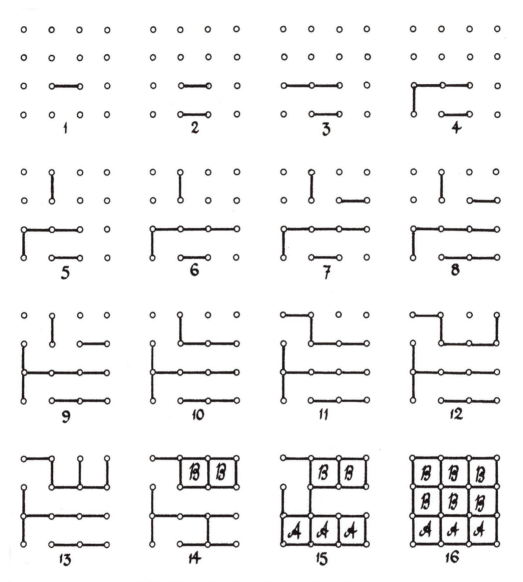

Figure 1. Arthur's and Bertha's First Game.

Bertha beats all her friends in this double-dealing way. Most children play at random unless they've looked quite hard and found that every move opens up some chain of boxes. Then they give the shortest chain away and get back the next shortest in return, and so on.

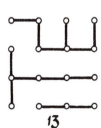 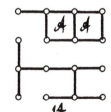 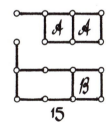 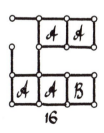

Figure 2. Bertha's Brilliance Astounds Arthur.

But when you open a long chain for Bertha, she may close it off with a double-dealing move which gives you the last 2 boxes but forces you to open the next chain for her (Fig. 3). In this way *she* keeps control right to the end of the game.

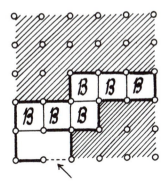

Figure 3. Bertha's Double-Dealing Move.

You can see in Fig. 4 just how effective this strategy can be. By politely rejecting two cakes on every plate but the last you offer her, Bertha helps herself to a resounding 19 to 6 victory. In the same position you'd have defeated the ordinary child 14 to 11.

Double-Dealing Leads to Double-Crosses

Each double-dealing move is followed, usually immediately, by a move in which two boxes are completed with a single stroke of the pen (Fig. 5). These moves are very important in the theory. We'll call them **doublecrossed** moves, because whoever makes them usually has been!

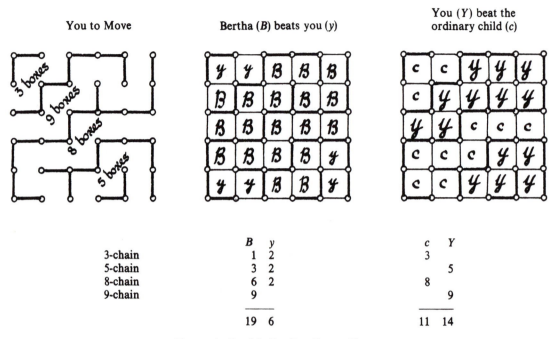

You to Move   **Bertha (B) beats you (y)**   **You (Y) beat the ordinary child (c)**

	B	y		c	Y
3-chain	1	2		3	
5-chain	3	2			5
8-chain	6	2		8	
9-chain	9				9
	19	6		11	14

Figure 4. Double-Dealing Pays off!

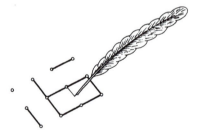

Figure 5. A Doublecross—Two Boxes at a Single Stroke.

Now Bertha's strategy suggests the following policy:

> Make sure there are long chains about
> and try to force your opponent to be
> the first to open one.

Try To Get Control . . .

We'll say that whoever can force her opponent to open a long chain has **control** of the game. Then:

> When you have control, make sure you
> keep it by politely declining 2 boxes
> of every long chain except the last.

... And Then Keep It.

The player who has control usually wins decisively when there are several long chains.

So the fight is really about control. How can you make sure of acquiring this valuable commodity? It depends on whether you're playing the odd- or even-numbered turns ... .

Figure 6. Which Is Dodie and which Is Evie?

Arthur and Bertha live next to the Parr family, in which there are two little sisters called Dodie and Evie (Bertha often teases them by calling them the Parrotty Girls!). You can see them playing the 4-box game in Fig. 6. Dodie's a year younger than Evie and so always has first turn in any game they play. They've got so used to playing like this that even when they're playing somebody else, Dodie always insists on taking the odd-numbered moves while Evie will only take the even-numbered ones:

> Dodie Parr: odd parity,
> Evie Parr: even parity.

The rule that helps them take control is:

Dodie tries to make the number of initial dots
+ doublecrossed moves *odd.*
Evie tries to make this number *even.*

Be SELFish about Dots + Doublecrosses!

In simple games the number of doublecrosses will be one less than the number of long chains and this rule becomes:

THE LONG CHAIN RULE

Try to make the number of initial dots
+ eventual long chains
even if your opponent is *Evie,*
odd if your opponent is *Dodie.*

The OPPOsite for Dots + Long Chains!

The reason for these rules is that whatever shape board you have on your paper, you'll find that:

	Number of dots you start with
+	Number of doublecrosses

| = | Total number of turns in the game. |

We'll show this in the Extras.

How Long Is "Long"?

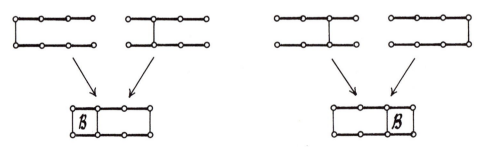

Figure 7. Bertha's Endgame Technique.

We can find the proper definition of long by thinking about Bertha's endgame technique. A **long chain** is one which contains 3 or more squares. This is because whichever edge Arthur draws in such a chain, Bertha can take all but 2 of the boxes in it, and complete her turn by drawing an edge which does not complete a box. Figure 7 shows this for the 3-square chain. A chain of 2 squares is *short* because our opponent might insert the *middle* edge, leaving us with no way of finishing our turn in the same chain. This is called (Fig. 8(a)) the **hard-hearted handout**.

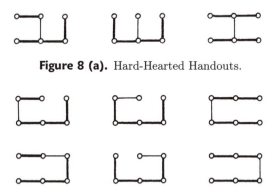

Figure 8 (a). Hard-Hearted Handouts.

Figure 8 (b). Half-Hearted Handouts.

When you think you are winning, but are forced to give away a pair of boxes, you should always make a hard-hearted handout, so that your opponent has no option but to accept. If you use a **half-hearted** one (Fig. 8(b)) he might reply with a double-dealing move and regain control. But if you're losing, you might try a half-hearted handout on the Enough Rope Principle (Chapter 1 Extras). Officially this is a bad move, since your opponent, if he has any sense, will grab both squares. But boys by billions, being bemused by Bertha's brilliance, blindly blunder both boxes back.

The 4-Box Game

When Dodie was *very* young, the girls often played the 4-box game and offset Dodie's first move advantage by calling it a win for Evie (the second player) when they each got 2 boxes:

> TWO TWOS IS A WIN TO TWO

At first Dodie would never give away a box if she could see something else to do, and Evie, who you can see is a very symmetrical player, would always win by copying Dodie's moves on the opposite side of the board. But after watching Bertha playing Evie, Dodie found how to counter this strategy by making a Greek gift on her 7th move. Evie can still win if Dodie dares to stray from the Path of Righteousness but must resist her temptation to make *every* move a symmetrical one (Fig. 9).

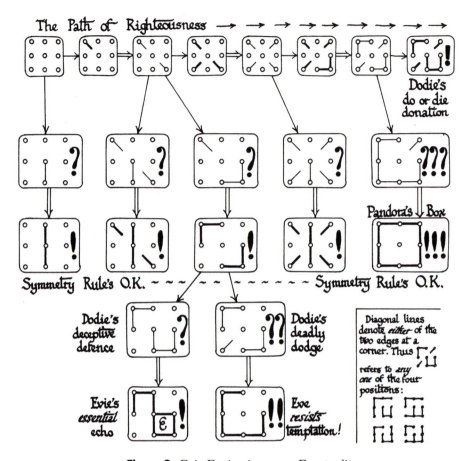

Figure 9. Evie Envisaging every Eventuality.

Even though Dodie has the win, it's much harder to write out in full her best plays against sufficiently cunning opponents. In Fig. 34 of the Extras we give an adequate strategy for Dodie and in Fig. 35 a complete list of $\mathcal{P}$-positions for both players. This little game is full of traps for the unwary, and those of you who have written to us for advice on becoming Professional Boxers will find these tables very useful in the bruising preliminary contests on the 4-box board. If the chain lengths are

$$
\begin{array}{lcl}
 & & \text{loop of 4} \\
4 & \text{or} & 2+2 \\
3+1 & & 1+1+1+1
\end{array}
$$

the winner will usually be

$$
\begin{array}{ccc}
\text{Dodie} & \text{or} & \text{Evie}
\end{array}
$$

in agreement with the Long Chain Rule, but on this small board, Dodie should often defy the rule and win by splitting the chains as $2 + 1 + 1$.

The 9-Box Game

Surprisingly, the Long Chain Rule makes the 9-box game seem easier than the 4-box one. This time Evie wins, and her basic strategy is to draw 4 spokes as in Fig. 10, forcing every long chain to go through the centre. Against most children this wins for Evie by at least 6-3, but Dodie can hold her down to 5-4, perhaps by sacrificing the centre square, after which Evie should abandon her spoke strategy. Of course, Evie's real aim is to arrange that there's just one long chain, and she often improves her score by forming this chain in some other way.

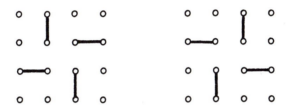

Figure 10. Lucky Charms Ward Off more than One Long Chain; Evie Puts Spokes in Dodie's Wheel.

Evie usually prefers to put her spokes in squares where another side is already drawn, and she's careful to draw spokes in only *one* of the two swastika patterns of Fig. 10. There usually aren't any double-crossed moves, so that Evie wins at the $(16 + 0 =)$ 16th turn.

Dodie tries to arrange *her* moves so that some spoke can only be inserted as a sacrifice, and *either* cuts up the chains as much as possible (maybe with a centre sacrifice) *or* forms *two* long chains when Evie isn't thinking. Every now and then a half-hearted handout has saved the game for her just when she thought that all was lost.

The 16-Box Game

We don't know who wins on the 4×4 box board, which makes a very interesting game to play. Evie tries to make the number of long chains 2, while Dodie tries to cut it down to 1 or force it up to 3.

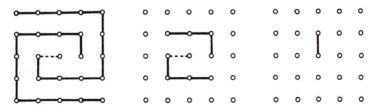

Figure 11. "Come into my symmetrical parlor!"

Evie beats many children with her symmetry strategy, but Dodie remembers her trick from the 4-box game. If she thinks her opponent will mimic her every move, she can lure him into the spider's web of Fig. 11(a), but when he's less predictable she finds it safer just to use the middle of the web (Fig. 11(b)). Dodie doesn't usually open with Fig. 11(c), because she finds the symmetry strategy very hard to beat.

Other Shapes of Board

To beat all your friends on larger square and rectangular boards you'll really need the Long Chain Rule. Remember to count a closed loop of 4 or more cells as *two* long chains and that each doublecross, no matter who makes it, changes the number of long chains you want. (Think of a doublecross as a long chain that's already been filled in.) It's good tactics to make the long chains as long as possible and avoid closed loops when you can, because you forfeit *four* boxes when declining a loop. These rules work for all large boards and even for triangular Dots-and-Boxes boards, like that in Fig. 12.

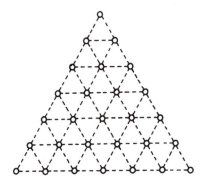

Figure 12. A Board with 28 Dots and 36 Triangular Cells.

Of course, if your opponent is also using the Long Chain Rule , the fight for control might be quite hard. The game of Nimstring, discussed in the rest of this chapter, is what control is all about. There's a piece in the Extras that describes some of the rare occasions when you might find it wise to lose control.

Dots-and-Boxes and Strings-and-Coins

You can play a dual form of Dots-and-Boxes, called **Strings-and-Coins**, with strings, coins and scissors. The ends of each piece of string are glued to two different coins or to a coin and the ground (each string has at most one end glued to the ground) and each player in turn cuts a new string. If your cut completely detaches a coin, you pocket it and must then cut another string (if there's one still uncut). The game ends when all coins are detached and the player who pockets the greater number is the winner.

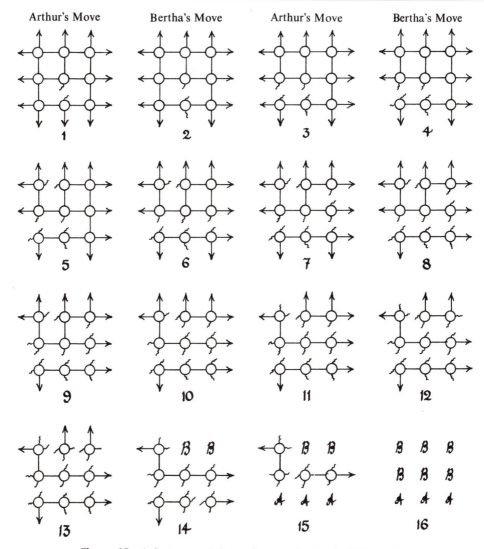

Figure 13. A Strings-and-Coins Game—the Dual of Figure 1.

Figure 13 shows the dual of Arthur's and Bertha's first game (compare it with Fig. 1). It started with 9 coins connected by 24 strings, 12 of them between coins and coins, the other 12 between coins and the ground. We use little arrows for strings that run to the ground. The coins and strings form the nodes and edges of a **graph**. It's easy to make a graph to correspond to any Dots-and-Boxes position. However, there are lots of graphs which *don't* correspond to such positions; for example the graph may have cycles of odd length or nodes with more than 4 edges, or the graph may be *non-planar*. In fact Strings-and-Coins is a generalization of Dots-and-Boxes.

Nimstring

The game of **Nimstring** is played on exactly the same kind of graphs as Strings-and-Coins, and you make exactly the same move by cutting a string (which is a *complimenting* move whenever you detach a coin). In Strings-and-Coins the winner is the player who detaches the larger number of coins, but Nimstring is played instead according to the Normal Play Rule. So, for ordinary Nimstring positions you *lose* when you detach the last coin, for then the rules require you to make a further move when it is impossible to do so. (But a Nimstring graph *may* have a string joining the ground to itself, and if the last move cuts *this* it doesn't detach a coin, and so *wins.*)

Nimstring looks quite different from Strings-and-Coins, but closer investigation shows that Nimstring is in fact a special case of Strings-and-Coins.

> You can't know all about Strings-and-Coins
> unless you know all about Nimstring!

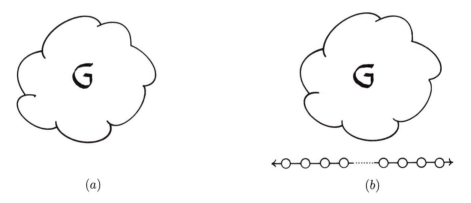

(a)   (b)

Figure 14. (a). A Hard Nimstring Problem. (b). This Strings-and-Coins Problem Is just as Hard.

Figure 14 shows the construction that proves this. If G represents an arbitrary Nimstring problem, we add a long chain to it and consider the resulting Strings-and-Coins game—the long chain should have more coins than G. Because the chain is so long and whoever first cuts a string of it allows his opponent to capture all the coins of the chain on his next turn, both players will try to avoid cutting any string of the chain. Neither player can force his opponent to move on the chain until all the strings of G have been cut. In other words, the only way to win the Strings-and-Coins game of Fig. 14(b) is to play a winning game of Nimstring on the graph G.

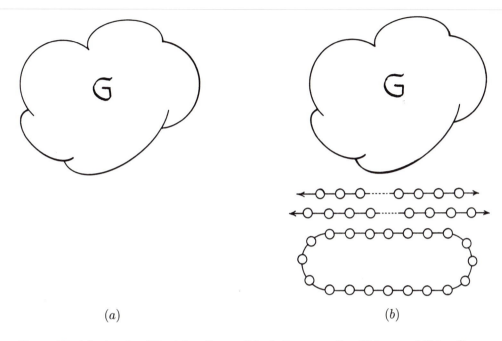

Figure 15. (a). Another Nimstring Game. (b). A Corresponding Strings-and-Coins Game.

Figure 15 shows another construction. This time we get the Strings-and-Coins game by adding several long chains and cycles to the Nimstring game G. If these are long enough the winning strategy for the Strings-and-Coins game is then:

> *If your opponent moves in G,*
> reply in G with a move from
> the winning Nimstring strategy.
> *If he moves in a long chain,*
> take *all but two* coins of that
> chain, leaving just the string
> which joins them.
> *If he moves in a long cycle,*
> take *all but four* coins of the
> cycle, leaving them as *two pairs*
> each joined by a string.

This strategy gives you all but 2 coins of each long chain and all but 4 of each long cycle, so it will win for you if the total number of nodes in the added chains and cycles exceeds

$$(\text{the number of nodes in } G)$$
$$+ \quad 4 \times (\text{the number of added long chains})$$
$$+ \quad 8 \times (\text{the number of added long cycles}).$$

In practice the Nimstring position will often contain (potential) long chains of its own, so that the strategy is of wider application. Recall that the "all but 2" principle was used by Bertha in her second game with Arthur (Fig. 2). Well-played games of Dots-and-Boxes are usually played like the corresponding Nimstring games, except at the very end. The last long chain in a Nimstring game is treated like any other; the winner takes all but the last 2 coins, which he gives to the loser by a hard-hearted handout. For the last chain in Dots-and-Boxes, of course, winner takes all!

Why Long Is Long

The argument explains why "long" must be defined precisely as follows. We should call a chain **long** if it contains 3 or more coins, because no matter which string of such a chain our opponent might cut, we may take all but 2 of its coins and finish by cutting another string of the chain. We must call a chain of 2 coins **short**, because he might cut the middle string and prevent us from declining those 2 vital coins (the hard-hearted handout). For a similar reason a closed loop of 2 or 3 coins would be called **short** (short loops don't arise in rectangular Dots-and-Boxes). However, a loop with at least 4 coins is called **long**, because we can politely decline the last 4 coins no matter which string our opponent cuts. Figure 16(a) shows how to do this on a 6-loop. When your opponent has cut the first string as shown, you only take 2 of the coins and then cut the string in the middle of the remaining 4. Figure 16(b) shows how this corresponds with Bertha's way of playing Dots-and-Boxes.

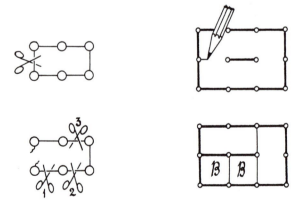

Figure 16. Bertha Politely Declines a Long Loop.

Well-played games of Dots-and-Boxes frequently lead to the duals of positions like those in Fig. 15(b). Most of the coins are in the long chains and loops, and the winner is whoever can force his opponent to cut the first string in one of these. It seems to be almost always the case that the winning strategy for Nimstring also gives the winning strategy for Strings-and-Coins. There are many other graphs than those satisfying the conditions of Fig. 15(0) for which this can be proved to happen. To win a game of Dots-and-Boxes or Strings-and-Coins, you should try to win the corresponding game of Nimstring and at the same time arrange that there are some fairly long chains about. In the rest of this chapter we'll teach you how to become an expert at Nimstring.

To Take or Not to Take a Coin in Nimstring

A coin which has only a single string attached is **capturable**. Whenever there's a capturable coin the next player has the option of removing the corresponding branch, thereby detaching the coin and getting another (complimentary) move. For some graphs this is the best move; for others, including one of those encountered by Bertha in the game of Fig. 2, the winning strategy is to refuse to detach the coin. As you might guess, the decision as to whether it's better to take a coin or decline it often depends on the entire graph. However a great deal can be deduced by examining only local properties of the graph near the capturable coin.

Any capturable coin must look like one of the six possibilities in Fig. 17. The string from the capturable coin goes either to the ground (Fig. 17(a)) or to another coin. If to another coin, the number of strings there is either one (Fig. 17(b)), two (Figs. 17(c), (e) and (f)) or three or

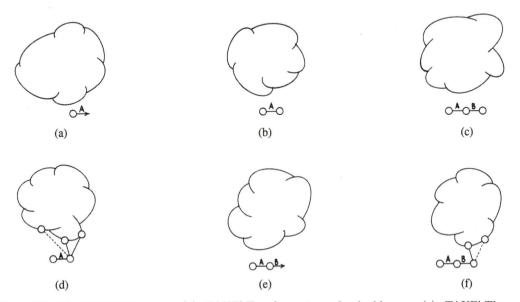

Figure 17. (a). TAKE! A free coin. (b). TAKE! Two free coins and a doublecross. (c). TAKE! Three free coins and a doublecross. (d). TAKE! A free coin. (e). WIN! (f). WIN! Half-hearted handouts.

more (Fig. 17(d)). If there are two strings, the second goes either to another capturable coin (Fig. 17(c)), or to the ground (Fig. 17(e)), or to a coin with two more or strings (Fig. 17(f)). In each of the six cases the cloud contains all the coins and strings not regarded as near enough to the capturable coin. The dotted lines in Figs. 17(d) and (f) are possible additional strings which may or may not be present.

We claim that in the first four cases (Figs. 17(a)-(d)) the player to move might as well cut string A and capture the coin, and in Fig. 17(c), he might as well continue by cutting string B, taking two more coins. For suppose you have a winning strategy starting from one of these graphs. If this tells you to complete your first turn by cutting only certain unlettered strings, then your opponent has the option of beginning *his* turn by cutting the lettered ones. But the same position will be reached if instead you first cut all the lettered strings and then cut the same unlettered ones as before. If there's any winning strategy at all, starting from these four cases, there's one which begins by cutting the lettered strings. So there's no loss in generality in supposing that a good player will TAKE a capturable coin of one of the four types in Figs. 17(a)–(d).

The other two positions (Figs. 17(e) and (f)) are much more interesting. If it's your turn to move in one of these two cases, you can either *detach* the capturable coin by cutting string A, or *decline* to take it by cutting string B. No matter what the rest of the graph might be, *one* or *other* of these two moves will WIN. But you might need to look at the whole graph to decide whether your winning strategy begins by cutting string A or string B!

This somewhat surprising result is proved by a cunning use of Strategy Stealing (Fig. 18).

We ask, for the games of Figs. 17(e) and (f):

who wins the smaller game G consisting of just the unlettered strings (Figs. 18(e) and 18(f))?

This is either the player who has to move from G or the player who doesn't. Whoever this fortunate player is, you should arrange to steal his strategy. If the player to move from G can win, then when playing from Fig. 17(e) or (f) you should start by cutting string A (which detaches a coin, so you continue), then cut string B (detaching another coin, so you continue again) and then begin the game on G, which of course you will play according to the winning strategy for the first player. On the other hand, if there's no winning move for the first player from G then, starting from Fig. 17(e) or (f), you should finish your turn immediately by cutting string B and so force your opponent to start the game G (he might as well start by cutting string A; if he doesn't, you will later).

The fact that the declining move forfeits 2 coins to your opponent makes no difference in Nimstring, where the winner is determined by the last move. In Strings-and-Coins (and Dots-and-Boxes) it *might* matter, but is unlikely to when there are long chains about.

Sprague-Grundy Theory for Nimstring Graphs

We now try to define values for arbitrary Nimstring graphs. We'd like these values to be nimbers so that we can use the ordinary Mex and Nim Addition Rules. The only trouble is that there are positions like that shown in Fig. 19.

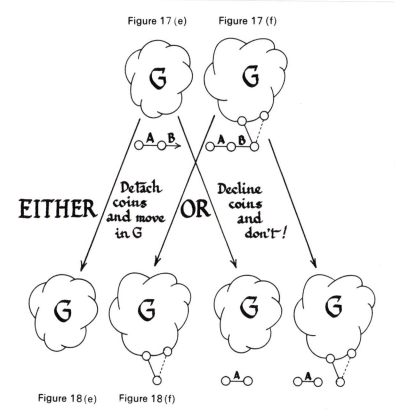

Figure 18. Strategy Stealing after a Half-hearted Handout.

Figure 19. A Loony Nimstring (or Dots-and-Boxes) Position.

Our discussion of Fig. 17(e) shows that no matter what graph G is added to this, the result is a win for the first player. The supposed value, $*x$, for Fig. 19 must therefore have the property that

$$*x + *y \neq 0$$

for every nimber $*y$, including even $*x$ itself, so that in particular

$$*x + *x \neq 0.$$

Those of you who have read Chapter 12 will see at once how to resolve this paradox. Figure 19 is what we call a **loony** position, whose value is 𝔇. The theory of complimenting and complimentary moves in that chapter applies to Nimstring (where the complimenting moves are those that capture coin's) and shows that every position has either an ordinary nimber value or the special value 𝔇. But don't reread Chapter 12 now, because we can easily summarize the rules for finding values at Nimstring:

The value of a graph without strings is 0.

The value of a graph with a capturable coin of one of the four types in Figs. 17(a)–(d) is equal to that of the subgraph obtained by removing the capturable coin(s) and its string(s).

The value of a graph with a capturable coin of one of the two types in Figs. 17(e) and (f) is 𝔇.

The value of a graph with no capturable coins is found from the values of the graphs left after cutting single strings by using the Mex Rule (Chapter 4).

VALUES FOR NIMSTRING

When adding these values, remember that

$$\mathfrak{D} + 0 = \mathfrak{D} + *1 = \mathfrak{D} + *2 = \ldots = \mathfrak{D} + \mathfrak{D} = \mathfrak{D},$$

as well as the ordinary nim-addition rules.

We show the calculation for some graphs in Fig. 20. When there are no capturable coins we write against each string the nim-value of the sub-graph obtained by cutting that string. Thus the last picture has options of nim-values 0, 1, 3, i.e. values *0, *1, *3, and so its own value is *2, because $2 = \text{mex}(0, 1, 3)$. Strings marked 𝔇 are loony options for the first player—if he cuts such a string he will LOSE against proper play even if some other graphs are added to the position. The nim-value of each graph is found from the mex of the numbers against its strings—in this you should ignore the 𝔇 values, which correspond to suiciding moves.

Although Dodie wins the 4-box game of Dots-and-Boxes, we can deduce from Fig. 20 that Evie wins the corresponding Nimstring position:

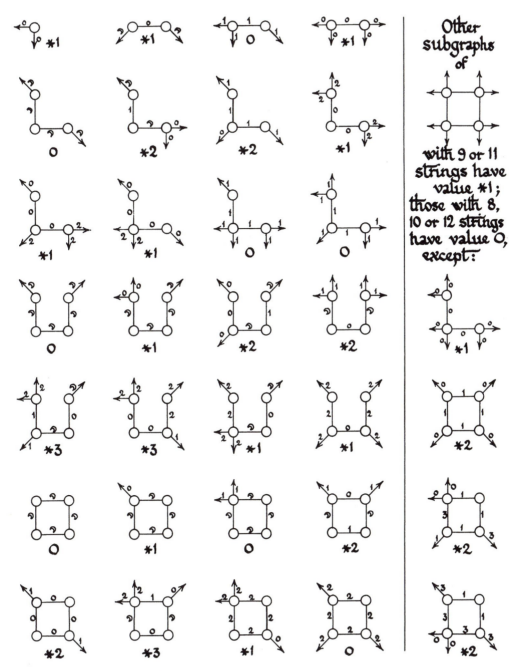

Figure 20. Working Out Values for Nimstring Graphs.

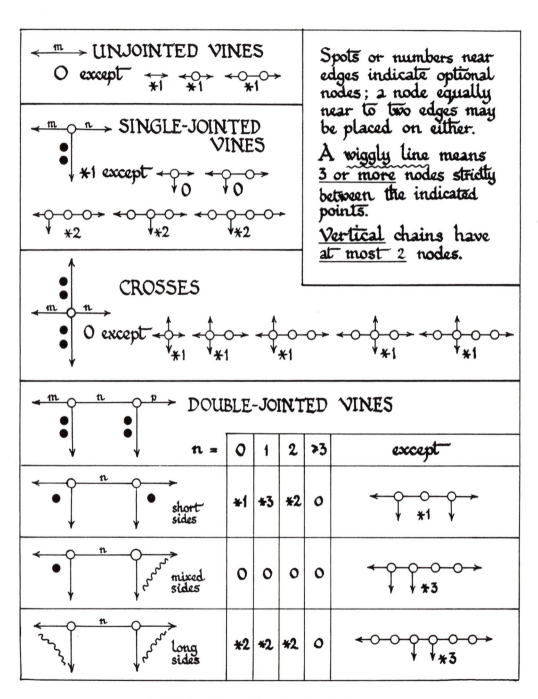

Figure 21. Noteworthy Nimstring Nimbers.

This means that even in Dots-and-Boxes Evie should win

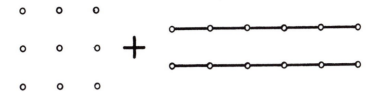

Figure 21 assembles answers for all graphs with at most 4 ends and no internal circuit. You'll find an extended table in the Extras that covers tree-like graphs with 5 ends.

All Long Chains Are the Same

Look at the various positions of Fig. 22, in which the clouds all conceal exactly the same thing, and the necklaces that hang from them all have at least three beads. The graphs all behave the same way in Nimstring because all the visible edges will always be loony moves.

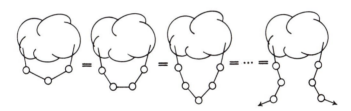

Figure 22. Three or More's a Crowd.

> Provided a chain has 3 or more nodes along it, the exact number doesn't make any difference to the value.

This makes it handy to have a special notation for long chains:

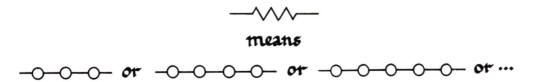

Which Mutations Are Harmless?

More generally we can put in or take out some beads on any Nimstring graph G to obtain **mutations** of the graph (a **bead**, of course, is a node with just 2 edges). Figure 23 shows a graph G and two mutations, H and K.

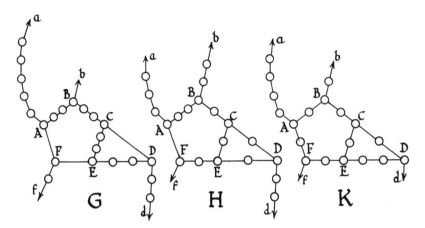

Figure 23. A Graph, a Harmless Mutation and a Killing One.

We'll use the word **stop** to mean *either* an arrowhead where the graph goes to ground (an **end**) *or* any of the nodes which have 3 or more edges (the **joints**). A path between two stops is **long** if it passes through 3 or more intermediate nodes, **short** otherwise. Mutation usually affects the value, but there are a lot of **harmless** mutations that don't:

A mutation between two graphs will certainly be harmless if every short path between stops in either graph corresponds to a short path in the other.

THE HARMLESS MUTATION THEOREM

In Fig. 23, H is a harmless mutation of G, since the only short paths are AE, Af, Ef, and the ones other than Aa that don't pass through a stop. But AE is long in K, and Cd is short, so this mutation is not covered by our theorem. In fact G and H have value $*2$, while K has value 0.

When G and H are related by a harmless mutation you just play H like G. A non-loony move must cut some string of a short chain between two points A and B that were stops at least until the move was made. A and B must have been stops in the original graph and we can find a similar non-loony move in the mutated graph because the distance between A and B will be short. (Fig. 24).

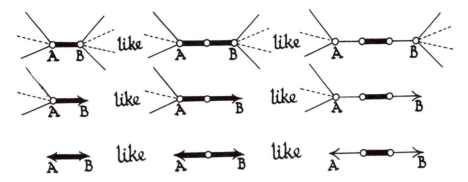

Figure 24. Like Moves in Harmless Mutations.

We can strengthen our Mutation Theorem a little:

> If the path between two stops
> passes through one end of a
> long chain, you needn't worry
> about the length of the path.

(For in a graph like Fig. 25—in which A or B might have been ends—AB won't become a chain unless someone makes a loony move cutting the long chain ending at C.)

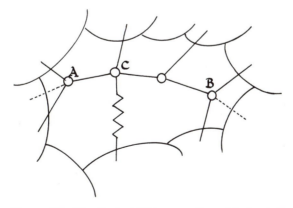

Figure 25. The Path AB Passes a Long Chain at C.

Chopping and Changing

There are lots of more drastic changes we can make to Nimstring graphs without affecting their values; for instance:

> Long chains snap!

This was hinted at in Fig. 22, and Fig. 26(a) shows how it's written in our long chain notation.

The remaining equivalences of Fig. 26 are more interesting. The middle equivalence of Fig. 26(b) is particularly useful (the left equivalence is a long chain that snaps). It asserts that when an edge runs to a node from which two long chains emanate, then this edge may be replaced by an edge running directly to the ground. More generally, when two long chains are attached to a node, all other edges ending at this node may be replaced by edges running to the ground (Fig. 26(c)).

The idea of the proof is that a node at the end of two long chains can't be captured until after someone concedes the game by making a loony move. We can apply the equality between the first and last parts of Fig. 26(b) to every branch that runs to the ground, so as to eliminate ground branches from every graph. But usually it's more convenient to use it in the other direction, eliminating many branches and nodes by introducing new ends. Sometimes, as in Fig. 26(d), this gives rise to a branch joining the ground to itself (a 0 by 1 game of Dots-and-Boxes!); such a branch contributes *1 to the value.

The equality between the first three parts of Fig. 26(e) follows from the Harmless Mutation Theorem, but that between these and the last three doesn't, because some short chains have become long. The letters label corresponding moves, and the 𝔇's show moves which we should ignore. Figures 26(b), (d), (f) and (g) show that we can sometimes eliminate circuits from our graphs—the last diagram of Fig. 26(f) is our shorthand notation for any of the previous three, which are harmless mutations of each other. Figure 26(h) has many variants, abbreviated in Fig. 26(i) (using the notation of Fig. 21).

Vines

A **vine** is a Nimstring graph without circuits or capturable nodes in which all the joints lie on a single long path (the **stem**) and each joint belongs to just 3 edges. The chain joining an end to its nearest joint is called a **tendril**, so a single-jointed vine has 3 tendrils (Fig. 21). Vines with more joints have 2 tendrils at their endmost joints and just 1 at intermediate ones. If the distance between two neighboring joints is long, the vine decomposes into two smaller ones because long chains snap, so we can suppose such distances *short*, if we like.

A **Twopins-vine** is one whose every distance between *non*-neighboring stops (which may be either ends or joints) is *long*. It is a remarkable fact that the value of any Twopins-vine is equal to that of a corresponding configuration in the game of Twopins (Chapter 15). Each joint with a *short* tendril becomes a column of *two* pins (even if it has also a long tendril); each other joint becomes a column of *one*; and two neighboring joints a long distance apart correspond to an *empty* Twopins column (Fig. 27). A bowling shot which removes a *single* column at Twopins corresponds to a *tendril* move at Nimstring; one which removes a *pair* of columns corresponds to a move on the *stem* of the vine.

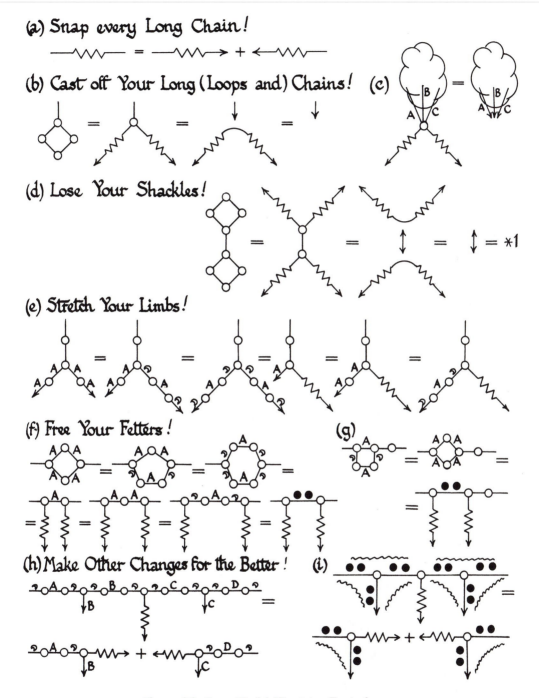

(a) Snap every Long Chain!

(b) Cast off Your Long (Loops and) Chains! (c)

(d) Lose Your Shackles!

(e) Stretch Your Limbs!

(f) Free Your Fetters! (g)

(h) Make Other Changes for the Better! (i)

Figure 26. Some Useful Nimstring Equivalences.

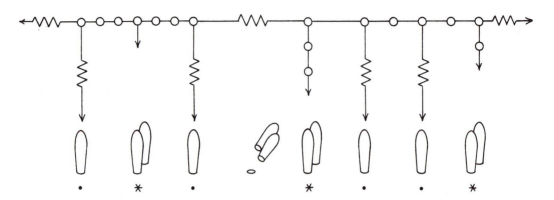

Figure 27. A Twopins-vine and a Game of Twopins.

Our remarks about vines show that:

> You can't know all about Nimstring
> without knowing all about Twopins!

If you've read Chapter 15 you'll know that Kayles and Dawson's Kayles are just special cases of Twopins, so, combining several slogans of this chapter:

> You can't know all about
> Dots-and-Boxes
> unless you know all about
> Kayles and Dawson's Kayles!

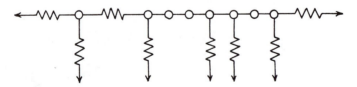

Figure 28. A Snappable Dawson's-vine, $D_1 + D_4$, Value $0 + *2 = *2$.

A **Dawson's-vine** (*Parthenocissus dawsonia*) is a Twopins-vine all of whose tendrils are long. Of course if any distance between neighboring joints is *long* the Dawson's-vine will snap, like the one in Fig. 28. If all these distances are *short*, the nim-values, D_n, of n-jointed Dawson's-vines are (Chapter 4):

n	0	1	2	3	4	5	6	7	8	9	11		13		15		17		19		21		23		25		27		29		31		33
D_n	0	0	1	1	2	0	3	1	1	0	3	3	2	2	4	0	5	**2**	**2**	3	3	0	1	1	3	0	2	1	1	0	4	5	**2**	7
D_{n+34}	4	**0**	1	1	2	0	3	1	1	0	3	3	2	2	4	4	5	5	**2**	3	3	0	1	1	3	0	2	1	1	0	4	5	3	7
D_{n+68}	4	8	1	1	2	0	3	1	1	0	3	3	2	2	4	4	5	5	9	3	3	0	1	1	3	0	2	…						

A *Kayles* position corresponds to a Twopins-vine with a *short* tendril at every joint. However, we can extend this class by observing that we don't need to worry about some of the distances between joints and ends:

> A vine is a **Kayles-vine** if
> (i) every joint has a *short* tendril, and
> (ii) every distance between two ends or two non-neighboring joints is *long*.

Again, if any distance between neighboring joints of your Kayles-vine is *long*, it snaps (Fig. 29). From Chapter 4, the nim-values, K_n, of unsnappable n-jointed Kayles-vines are:

n	0	1	2	3	4	5	6	7	8	9	10	11	12	13	14	15	16	17	18	19	20	21	22	23
K_n	0	1	2	**3**	1	4	**3**	2	1	4	2	**6**	4	1	2	**7**	1	4	**3**	2	1	4	**6**	7
K_{n+24}	4	1	2	8	**5**	4	7	2	1	8	**6**	7	4	1	2	**3**	1	4	7	2	1	8	2	7
K_{n+48}	4	1	2	8	1	4	7	2	1	**4**	2	7	4	1	2	8	1	4	7	2	1	8	**6**	7
K_{n+72}	4	1	2	8	1	4	7	2	1	8	2	7	4	1	2	8	1	4	7	2	1	8	2	7

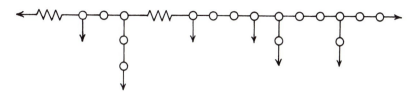

Figure 29. A Snappable Kayles-vine, $K_2 + K_4$, Value $*2 + *1 = *3$.

The correspondence between the Twopins-vines and the game of Twopins enables us to interpret the Decomposition Theorem of Figs. 26(h) and (i) as a generalization of the Decomposition Theorem for Twopins. There are Nimstring generalizations for all the Twopins equivalences:

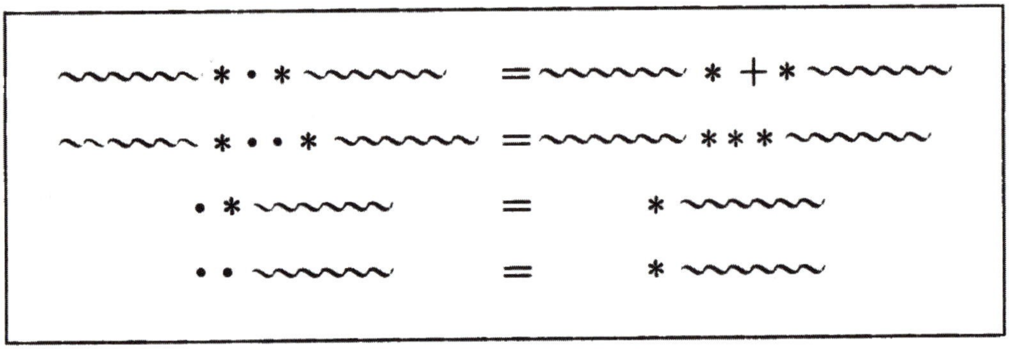

~~~~~ * • * ~~~~~~~ =~~~~~~ * + * ~~~~~

~~~~~ * • • * ~~~~~ =~~~~~ * * * ~~~~~

• * ~~~~~   =   * ~~~~~

• • ~~~~~   =   * ~~~~~

The last two enable us to suppose that the endmost joints of a Twopins-vine have short tendrils (they correspond to uses of Fig. 26(b)). Collectively the Twopins equivalences allow us to suppose that the joints with no short tendrils come in strictly internal blocks of 3 or more and so all the simplest Twopins-vines reduce to compounds of Kayles-vines. Figure 30 is a small Twopins dictionary culled from Chapters 4 and 15. The equivalences of Fig. 26 enable us to show that many graphs that don't look like vines are really equivalent to them; for instance Fig. 31(a) is equivalent to D_8.

Kayles-vines, K_n		n	Dawson's-vines, D_n				Other Twopins-vines	
*	= *1	1	·	=		= 0	* * · · · *	= *2
* *	= *2	2	· ·	=	*	= *1	* * * · · · *	= *3
* * *	= *3	3	· · ·	=	* ·	= *1	* * · · · * *	= *1
* * * *	= *1	4	· · · ·	=	* *	= *2	* * · · · · *	= *4
* * * * *	= *4	5	· · · · ·	=	* + *	= 0	* * * * · · · *	= *5
* * * * * *	= *3	6	· · · · · ·	=	* * *	= *3	* * * · · · * *	= *4
* * * * * * *	= *2	7	· · · · · · ·	=	* · · · *	= *1	* * * · · · · *	= *3
* * * * * * * *	= *1	8	· · · · · · · ·	=	* · · · · *	= *1	* * · · · · * *	= *3
* * * * * * * * *	= *4	9	· · · · · · · · ·	=	* · · · · · *	= 0	* * · · · · · *	= *3
* * * * * * * * * *	= *2	10	· · · · · · · · · ·	=	* · · · · · · *	= *3	* · · · * · · · *	= *1

Figure 30. Various Vines Values.

Twopins-vines are **decomposing** in the sense that when any branch of the vine is removed the new vine decomposes—often by snapping a chain—into two smaller ones. Some other vines, including that of Fig. 31(b), are decomposing in the same sense. It is rather straightforward to compute the value of a decomposing vine from the values of those of its subvines which include all of a consecutive sequence of the original tendrils. Since the number of such subvines is proportional only to the square of the number of tendrils this idea is feasible for quite long decomposing vines, and can easily be implemented on a computer.

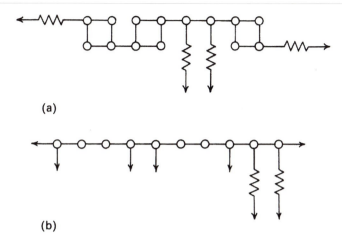

(a)

(b)

Figure 31. Two Uses of Figure 26. (a) A graph equivalent to the Dawson's-vine D_8. (b) A decomposing vine.

Dots-and-Boxes is, like other good games, remarkable in that it can be played on several different levels of sophistication.

- First, there's Arthur's way: you just don't open up any boxes unless you have to and then you open as few as possible. This seems to be the only level that many players reach.

- Then there's Bertha's double-dealing endgame technique which gets the winner a lot of boxes at the finish and makes it seem likely that Nimstring will be useful.

- Next comes the Parrotty girls' parity rule for long chains.

- Then we realize that to get the right parity is an exercise in Sprague-Grundy theory, so we need tables of nim-values.

- The unwieldiness of these tables forces us to use equivalence theorems whenever we can, and to look for interesting classes of analyzable graphs.

- We can use Twopins theory to reduce many games to positions in the well-known games of Kayles and Dawson's Kayles.

- Finally, experts will need to know something about the rare occasions when the Nimstring theory does not give the correct Dots-and-Boxes winner.

What moves would you recommend for the positions in Fig. 32? We give our own recommendations in the Extras.

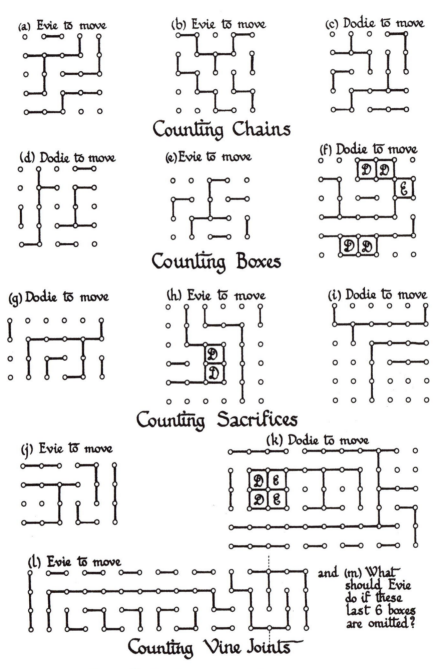

Figure 32. Try These Dots-and-Boxes Problems.

Extras

Dots + Doublecrosses = Turns

Suppose we play a Dots-and-Boxes game, starting with D dots, that takes T turns to draw L lines and finish with B boxes. Then if there are no doublecrosses, each line except the last either creates just one box or hands the turn to the next player, so

$$L = B + T - 1.$$

However Fig. 33 shows that

$$L = B + D - 1,$$

so a game with no doublecrosses lasts for exactly the same number of turns as the initial number of dots. But each doublecross creates 2 boxes instead of 1, so in general the number of turns will be the number of dots we started with plus the number of doublecrosses.

When we've broken **B** edges to flood the **B** boxes ...

...there'll be just $D - 1$ roads, all leading to Rome (1 from each other town).

Figure 33. Euler via Rademacher and Toeplitz.

How Dodie Can Win the 4-Box Game

Figure 34 shows a sufficient set of $\mathcal{P}$-positions to enable Dodie to win the 4-box game. Figure 35 shows *all* $\mathcal{P}$-positions except those in which a player has already signed enough boxes to win, classified according to the number of moves made. Figure 36 shows the three $\mathcal{N}$-positions in which a sacrifice wins but a non-sacrifice *loses*. In these figures broken lines indicate boxes with three sides already drawn.

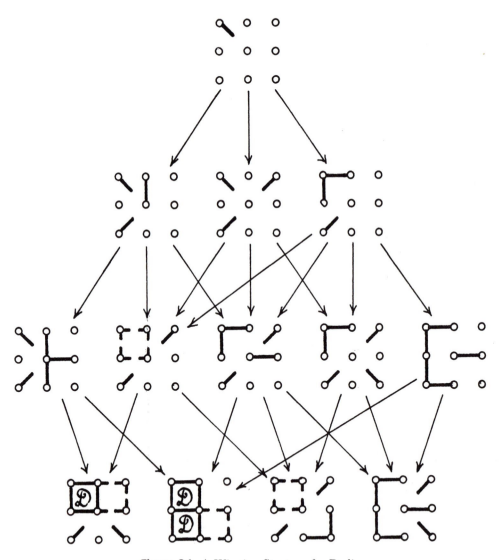

Figure 34. A Winning Strategy for Dodie.

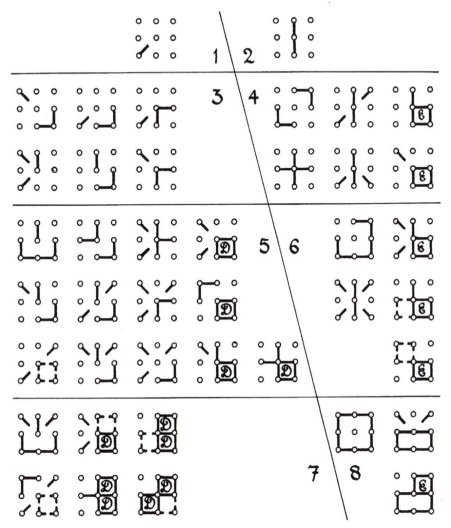

Figure 35. All $\mathcal{P}$-positions in the 4-box Game.

Figure 36. Three Tricky $\mathcal{N}$-positions.

When Is it Best to Lose Control?

It is clearly not *always* a good idea to keep control. For suppose all that's left of a game is 1001 chains, all of length 3, and your opponent has just opened the first of these. If you slavishly insist on keeping control to the very end, you'll have to give away 2 boxes in each chain but the last and so you'll only get 1003 to your opponent's 2000.

What the question really boils down to is this: when your opponent has just opened a long chain or made a half-hearted handout, should you, like Bertha, decline the last 2 boxes, or, like Arthur, grab them all and be forced to move elsewhere? Suppose for a moment that you use Bertha's tactic and give up 2 boxes in order to force your opponent to move first in the rest of the position, in which, by playing perfectly, you get D more boxes than your opponent. Then Arthur's strategy would take those 2 boxes and give you D less of the rest than your opponent. Comparing:

Bertha's technique	Arthur's technique
$-2 + D$	$2 - D$

shows you should

> adopt Bertha's technique
> unless D is less than 2.
> (Either will do, if D is just 2.)

That's all very well, but you still won't know which is best if you don't know the value of D. We can't say much about this in general, but we have a rule which gives D when the position is made up entirely of long chains of lengths

$$a, b, c, \ldots .$$

For such a position

$$D = (a - 4) + (b - 4) + (c - 4) + \ldots + 4$$

provided the right side is positive; otherwise $D = 1$ or 2.

Using this rule we can answer our question for positions made entirely of chains:

> You should keep or gain control, *unless*
> there are *evenly* many short chains, *and*
> *either* there are no long chains
> *or* the long chains can be partitioned
> into two sets, each of average
> length strictly less than 4.

The A.B.C. of Control when Long Chains are Short.

Of course, for such positions, keeping or gaining control involves making Bertha's move if this would leave an even number of unopened short chains, and Arthur's move otherwise.

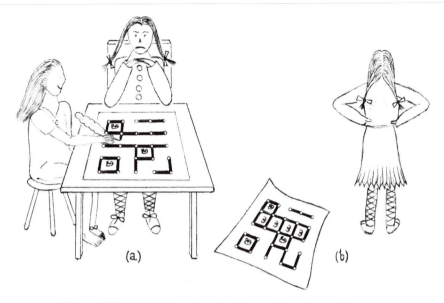

Figure 37. Dodie's Drawing the Game—Evie Loses Control.

Figure 37(a) shows a position where Evie, who's not feeling very well, has only managed to keep control so far by sacrificing 3 boxes. Dodie's now foolishly opening a chain of length 4 with the move shown. Since the remaining long chains satisfy the exceptional condition in our A.B.C., Evie's best move (Fig. 37(b)) is to *lose* control and take all 4 squares of the chain as Arthur would. The boxes are then divided:

		4-chain	three 3-chains			
to Evie		4	2	2		an $\frac{8}{8}$ tie,
to Dodie	3 already +	1	1	3		

where Bertha's response would give

to Evie	2		1		1	3		$\frac{7}{9}$ loss.
to Dodie	3 already +	2	2	2				

Computing the Values of Vines

On most graphs made of Twopins-vines and long chains a winning Nimstring strategy really does win at Dots-and-Boxes. For let V be the number of separate vines, counting long chains as unjointed vines. *Don't* use Fig. 26(a) to decompose vines with long stems—*instead* let I denote the number of these internal long stems. Let J be the total number of joints and L the Nimstring loser's present score. We study the quantities

$$f = L + 2J + 2V, \quad g = I + 2L + 3J + 4V$$

during any double turn, consisting of the loser's move and the winner's reply, assigning any boxes given away by a move to that move rather than to the next one. First for the Nimstring winner's move:

Move on	Change in					
	I	L	J	V	f	g
Stem	0	≤ 2	-2	1	≤ 0	≤ 2
Inner tendril	≤ 1	≤ 2	-1	0	≤ 0	≤ 2
End tendril	≤ 0	≤ 2	-1	0	≤ 0	≤ 1

And now for the Nimstring loser's move:

Move on	Change in						and including winner's reply	
	I	L	J	V	f	g	f	g
Stem	0	0	-2	1	-2	-2		
Inner tendril	≤ 1	0	-1	0	-2	≤ -2	≤ -2	≤ 0
End tendril	≤ 0	0	-1	0	-2	≤ -3		
Loony stem move that winner must accept	≤ 0	0	-2	1	-2	≤ -2	≤ -2	≤ 0
Loony tendril move that winner must accept	≤ 1	0	-1	0	-2	≤ -2	≤ -2	≤ 0
Loony chain move (declined by winner)	0	2	0	-1	0	0	0	≤ 0
Loony stem move (declined by winner)	≤ 0	2	-2	1	0	≤ 2	0	$\le 2^\dagger$
Loony tendril move (declined by winner)	≤ 1	2	-1	0	0	≤ 2	0	$\le 2^\dagger$

(No loony chain move makes the winner *accept*, because by *declining* he leaves the value unchanged.) The next to last column shows that f *never* increases and so:

> If the number of nodes, N,
> in the game exceeds
> $$4(J+V)$$
> then the Nimstring winner
> wins the Dots-and-Boxes game.

(For the loser's score at the end of the game will be less than $N/2$.

Since all Dawson's-vines and many Twopins-vines have more than four nodes per joint, they satisfy this condition. *If g never increased*, we could similarly assert that the Nimstring strategy works for Twopins-vines with

$$N > I + 3J + 4V,$$

but since the daggered entries can be positive a skilful Nimstring loser might win the occasional Dots-and-Boxes game.

Such cases are very rare. The Nimstring loser can increase g only by choosing a loony stem or tendril move that the Nimstring winner must decline (*most* loony moves can be accepted). Usually the winner has other opportunities to decrease g by playing on an end tendril or making a move conceding fewer than two boxes.

Even though we have been able to construct some examples (Fig. 32(m)), the difficulties of composing them and the closeness of their scores reinforce our opinion that:

> Your best chances at
> Dots-and-Boxes
> are likely to be found
> by the Nimstring strategy.

Loony Endgames Are NP-Hard

If you're faced with a position in which all the edges are on long chains you'll lose at Nimstring because only loony moves are possible. But if you've already got lots of boxes you might still manage to win the Dots-and-Boxes game. How do you find *which* loony move to make to stop your opponent from catching up?

To simplify the argument we'll suppose that the last move will take place on a chain (between ground and ground) that's long enough to ensure that your opponent's best strategy for the remaining boxes is the Nimstring strategy, which requires that he conclude each turn, except the last, with a double-dealing move. Any of the m moves you make on isolated cycles will give you 4 boxes, while any of the other n moves on chains (except the last) will give you 2 boxes each, so your score will be

$$4m + 2n - 2.$$

Suppose the graph has j joints with total valence v, counting grounded ends as having valence 1 each. A move on an isolated cycle doesn't change the valence, but a move on a chain decreases the valence by 1 at each end, except that whenever the valence of a joint changes from 3 to 2 that joint disappears. This happens just once for each joint, so

$$v = 2n + 2j$$

and your score will be

$$4m + v - 2j - 2.$$

Since v and j are fixed, we want to make as many moves on isolated cycles as possible. These isolated cycles are disjoint, and any disjoint set of cycles can be isolated just by playing all chain moves first.

> You can't know all about (possibly
> generalized) Dots-and-Boxes unless
> you know all about how to find
> a largest set of node-disjoint cycles
> in an arbitrary (possibly non-planar)
> graph.

Finding a largest set of node-disjoint cycles in arbitrary graphs is known to be NP-hard (see the Extras to Chapter 7).

In practice, most Dots and Boxes games are played on reasonably small boards, rarely larger than 10×10. In all loony positions that we have ever encountered on these small boards, we have easily found a maximal set of node-disjoint loops, by hand, in under a minute. But finding an order to play the loony moves which forces your opponent to retain control can still be a challenging problem, which is addressed in a paper by Berlekamp and Scott.

Solutions to Dots-and-Boxes Problems

Here are *our* answers to the problems in Fig. 32.

(a) Evie wants an even number of long chains. She immediately establishes just two by drawing either edge at the top left-hand corner.

(b) This time Evie establishes two long chains by sacrificing the box whose lower left corner is the central dot. An additional sacrifice in the lower left-hand corner will be needed if Dodie tries to make a third long chain there.

(c) Dodie wants an odd number of long chains. She should sacrifice two boxes by a hard-hearted handout drawn rightwards from the central dot, and will win 9–7.

(d) Neither player can afford to sacrifice four boxes of the central loop, so the long chain theory doesn't really apply. Either player can force a tie (making no sacrifices). A well-played endgame will have chains of length 3 at top and bottom and a loop of 4 in the centre. The left side may be a single chain of length 4 or a pair of chains of lengths 1 and 3; either position is a tie.

(e) A little trap for Nimstring players! The *dotted* move is the only good Nimstring move, but involves too much sacrifice and will lose the Dots-and-Boxes game 5–7 against an extremely skilful opponent. The *dashed* move loses at Nimstring, but only if the opponent sacrifices two boxes on the next turn, and with the board then broken into many small pieces we get a tie.

(f) The Nimstring game is essentially over, but at Dots-and-Boxes what matters is whether the top left-hand corner becomes a chain of length 2 or two chains of length 1. Dodie should force the former by drawing the second edge on the top row and will win 13–12 rather than losing 12–13.

(g) Dodie forces a treble sacrifice! She wants an even number of long chains but can see only one. Her dotted opening move threatens to create a second long chain at EFG on the next move. Since Evie can be prevented from making a third long chain, she must cut between F and G, sacrificing two boxes. Accepting these, Dodie repeats her threat by the dashed move, which threatens a long chain at CDE, forcing Evie to sacrifice D and E. Dodie accepts these and repeats the threat yet again, drawing the left edge of B. Although Evie wins the Nimstring game by sacrificing B and C, Dodie will have 8 boxes: F, G; D, E; B, C; N, O; enough to win the Dots-and-Boxes game.

(h) Evie does likewise! Starting with the next to last of the top edges she repeatedly threatens to construct a third long chain at the right. Dodie can stop this only by three 2-box sacrifices. Evie then makes a loony move conceding the chain of length 3 to the left of the 2 captured boxes, acquiring 2 boxes after the resulting doublecross. Since she has now changed sides, she threatens to build another long chain in the top left-hand corner, forcing a further sacrifice from Dodie. She then stops further growth of the length 7 chain and awaits her last 3 boxes to win 13–12.

(i) A very complicated position! Dodie *must* prevent a third long chain from forming in the top row. She first sacrifices one of the top corner boxes (and will probably need more sacrifices) and strives vigilantly to chew up as much of the empty space as she can by extending her long chains. If Evie sacrifices either long chain too soon, Dodie accepts.

(j) An easy one! There is a treble-jointed Kayles-vine, value $*3$, and four boxes at the lower left, value $*2$ from Fig. 20. Evie wins by drawing the middle top edge or rightmost bottom one, which reduces K_3 to K_2. (There are other moves that do this, but they sacrifice too much.)

(k) There's a 4-jointed Kayles-vine, value $*1$, at the bottom, and four boxes at top right, worth $*3$ from Fig. 20. The rest of the figure is a 5-jointed Kayles-vine, value $*4$, under a disguise you can strip off by looking at Figs. 26(a) and (b). Dodie's Nimstring move must therefore replace K_5 by $K_3 + K_1$ which she can do only by drawing the vertical edge at the top left-hand corner or by isolating the loop just to the right of the captured boxes. In this problem it's only if Dodie plays carefully that her Nimstring strategy will also win at Dots-and-Boxes. When she has a choice of several Nimstring moves, she should select whichever scores more boxes. On the Kayles-vines any stem move (by either player) leads ultimately to another long chain, which will give two more boxes to Evie. So Dodie prefers to make tendril moves whenever possible, while Evie selects stem moves which make Dodie respond with more stem moves.

(l) A unique winning move! Cut the 12-jointed Kayles-vine into two 5-jointed ones by separating N from O. The rest is easy!

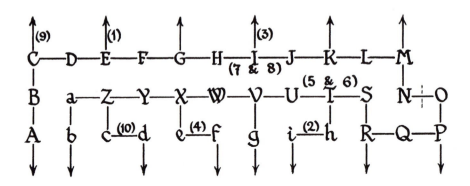

(m) In the modified problem, if Evie really respected Dodie's skill, she would resign! If she opened as before by sacrificing N, O, Dodie could unground E. Evie is then faced with $K_3 + K_1 + K_5$, and the Nimstring replies give Dodie two more boxes, e, f or h, i, say the latter (2). Dodie then (3) ungrounds I, leaving $4K_1 + K_3$, and Evie can't do better than (4) giving Dodie e and f (say) too. Evie now has $6K_1$ and has won the Nimstring game, but Dodie makes a loony move (5) on STU, which Evie's sadly forced to decline (6), conceding two more boxes. Dodie can make another loony move (7) on HIJ and collect two of those as well (8). Finally (9), Dodie can reduce $2K_1$ to K_1 by ungrounding C, which makes Evie (10) sacrifice c, d (or a, b). The resulting position

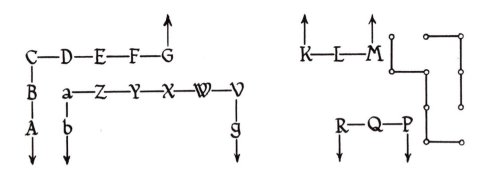

has five chains, lengths 3, 3, 4, 7, 8, but Evie has only 2 boxes to Dodie's 12. Although of the remaining chains Evie collects 17 boxes to Dodie's 8, Dodie wins 20 to 19! Rather than risk this disgrace, we recommend to Evie a timid opening such as ungrounding E; it's just conceivable that Dodie doesn't remember the first nine Kayles values!

Some More Nimstring Values

As in Fig. 21, dots near edges are optional additional nodes and a dot equally near two edges may be placed on either. A wiggly line means 3 or more nodes strictly between the indicated points. The symbol

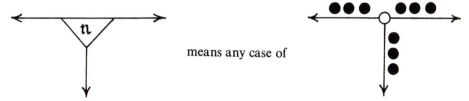

means any case of

that has value $*n$ (see Fig. 21).

Nimbers for Nimstring Arrays

To show the sides on which rectangular arrays are grounded we give their dimensions with primes (or double primes), for example:

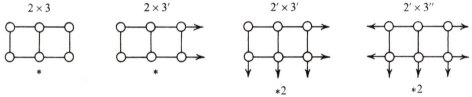

Tables 1 and 2 give values for such rectangular arrays:

	$1'$	2	$2'$	3	$3'$	4	$4'$	5	$5'$	6	$6'$	7	$7'$	8
2	$*$	0	$*2$	$*$	$*$	0	$*2$	$*$	$*3$	0	$*2$	$*$	$*3$	0
3	$*2$	$*$	$*2$	$*$	$*2$	$*$	$*2$	$*$						

Table 1. Nimstring Values for Ungrounded Rectangular Arrays or Arrays Grounded along One Edge.

n	2	3	4	5	6	7	8	9	10	11
$1' \times n$	$*$	$*2$	$*$	$*2$	$*3$	0	$*3$	0	$*$	$*2$
$1' \times n'$	0	$*$	0	$*3$	$*2$	$*3$	$*2$	$*5$	$*4$	$*5$
$1' \times n''$	$*$	0	$*$	0	$*$	$*2$	$*3$	$*$	$*3$	
$2' \times n$	$*2$	$*2$	$*$	$*$	$*$					
$2' \times n'$	$*2$	$*2$	$*$	$*$						
$2' \times n''$	0	$*2$	$*5$	$*$						

Table 2. Nimstring Values for Arrays Grounded at 1, 2 or 3 Edges.

Figure 39 shows Nimstring values for some less regular arrays.

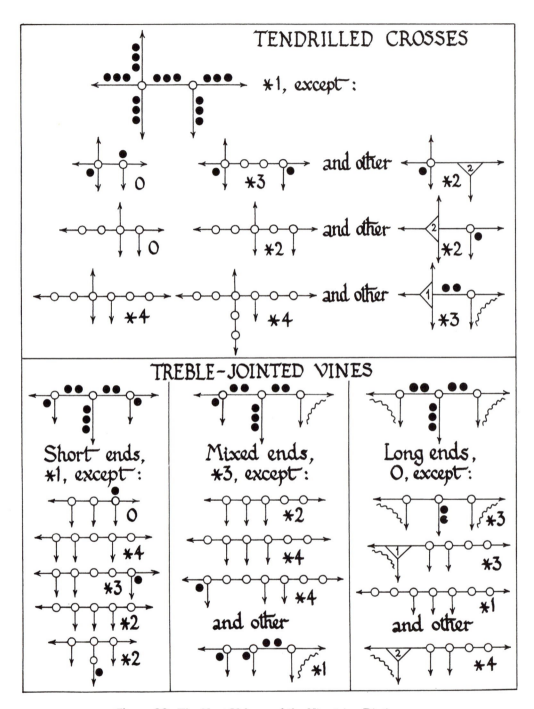

Figure 38. The Next Volume of the Nimstring Dictionary.

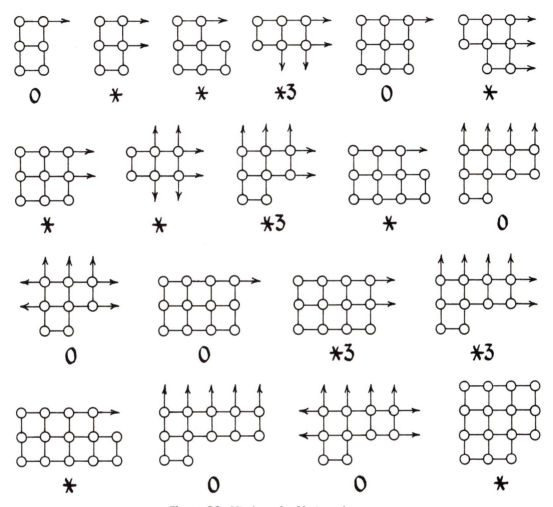

Figure 39. Nimbers for Various Arrays.

We have seen that we can play Nimstring on any graphs; here are the nim-values of small complete graphs, K_n and complete bipartite graphs, $K_{m,n}$:

n	2	3	4	5	6	7	8	9	10
K_n	0	*	0	*	*				
$K_{2,n}$	0	*	0	*	0	*	0	*	0
$K_{3,n}$	*	*	*	*	*	*			
$K_{4,n}$	0	*	0	*					

Does the value of $K_{m,n}$ depend only on the parity of $(m-1)(n-1)$?

References and Further Reading

Several additional results on Dots-and-Boxes can be found in Berlekamp's booklet, which includes over 100 problems and solutions. More complete and accurate solutions to some of the problems presented in the first edition of that booklet may be found at http://www.cae.wisc.edu/~dwilson/boxes/.

Elwyn Berlekamp, *The Dots and Boxes Game: Sophisticated Child's Play*, A K Peters, Ltd, Natick, MA, 2000; MR 2001i:00005.

Elwyn Berlekamp and Katherine Scott, Forcing your opponent to stay in control of a loony Dots-and-Boxes endgame, in Richard Nowakowski (ed.) *More Games of No Chance*, (Berkeley CA 2000) Math. Sci. Res. Inst. Publ., 42 (2002) Cambridge Univ. Press, Cambridge, UK, 317–330.

John C. Holladay, A note on the game of dots, *Amer. Math. Monthly*, **73** (1966) 717–720: M.R.

Hans Rademacher and Otto Toeplitz, *The Enjoyment of Mathematics*, Princeton University Press, 1957. Pages 75–76 give the proof of Euler's theorem.

Katherine Scott, Loony endgames in dots and boxes, MSc. thesis, Univ. of California, Berkeley, 2000.

Julian West, Championship-level play of Dots-and-Boxes, in Richard Nowakowski (ed.) *Games of No Chance*, (Berkeley CA 1994) Math. Sci. Res. Inst. Publ., 29 (1996) Cambridge Univ. Press, Cambridge, UK, 79–84.

-17-

Spots and Sprouts

He shall not live, with a spot I damn him.
William Shakespeare, Julius Caesar IV, i, 6.

The games we treat here are played with spots (or crosses) on a piece of paper, the move being to join two spots by a curve satisfying various conditions specified in the rules of the game. *We shall always demand that no curve crosses itself or another curve.* We have just devoted a whole chapter to such a game, but here we shall consider games whose theories, while not all trivial (or even all complete) will occupy only a few pages each. We had to make an exception for Lucasta, with whom we fell in love.

Rims

Here the move is simply to draw a loop passing through at least one and arbitrarily many of the spots. The only further condition is that no two loops may cross. A typical Rims position is shown in Fig. 1. What should be our next move, supposing normal play?

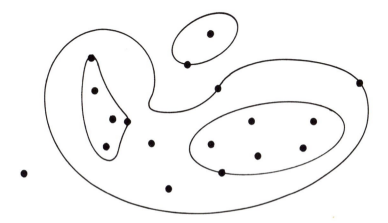

Figure 1. A Game of Rims (or Rails).

On examining the position, we see that the loops divide the plane into regions containing respectively 5, 2, 3, 1, 1 spots (and sometimes other regions with internal spots). When we make a move in a region with n spots, we automatically divide it into two regions with a and b spots, where $a + b$ is less than n, but a and b are otherwise arbitrary. It follows that Rims is merely a disguised form of Nim with the additional possibility of dividing the heap we've just reduced into two smaller ones. This is the game $0 \cdot \dot{7}$ in the octal notation of Chapter 4, where we saw that the extra possibility doesn't affect the strategy, so the only good move in Fig. 1 is to draw a loop through just 4 of the spots in the 5-spot region. The theory of Misère Nim tells us that exactly the same move should be made in Misère Rims. In general we move so that the nim-sum of the spot counts is zero, except that in the misère form we must make the nim-sum 1 if every spot count is 0 or 1.

Rails

Let us require that the loop must pass through just one or two spots, the rules otherwise being as in Rims. What should now be our move in Fig. 1? The legitimate moves in an n-spot region produce regions of a and b spots, where we require that $a + b = n - 1$ or $n - 2$. Since these moves correspond exactly to the legal moves in Kayles, our moves can be deduced from the theory of that game. For the Rails position of Fig. 1, this tells us to draw a loop through just one of the spots in the 5-spot region, in either normal or misère play.

Many other octal games can be reformulated very nicely as spot and loop games, and we find by observation that more people can be persuaded to play them this way. Often the geometrical form suggests particular rules very naturally, and sometimes the rules suggested do not quite correspond to natural games with heaps. Here are two further examples.

Loops-and-Branches

The move is to join two spots together, *or* join a single spot to itself so as to form a loop. No spot may be involved in two different moves. The game is isomorphic to the octal game $\cdot\mathbf{73}$, for which we computed the nim-values in Chapter 4 (Table 6) and the reduced forms in Chapter 13 (Extras, Table 5, Notes A and T, Adders). The patterns in Table 1 continue indefinitely.

n	0	1	2	3	4	5	6	7	8	9	...
nim-value	0	1	2	3	0	1	2	3	0	1	...
reduced form	0	1	2	3	2+2	3+2	2+2+2	3+2+2	2+2+2+2	3+2+2+2	...

Table 1. Nim-values and Reduced Forms for Loops-and-Branches.

So we have complete strategies in both normal and misère play. In both cases we move so that the nim-sum of the nim-values is zero, *except* that in misère play we must make the nim-sum one if every region has at most one spot.

Contours

This game is rather more interesting. The move is to draw a closed loop (or *contour*) through just one spot, with the side condition that every loop must have at least one spot strictly inside (possibly internal to some further contours). In other words, when we view the position as a system of contours drawn on a map, every hill must have its peak marked (and every valley its bottom).

In this game, we must distinguish a region containing n spots and nothing else (type n) from one which, in addition to n free spots, contains a contour or contours with their internal spots (type $\hat{n}$). But the number or structure of the contours within a region of type $\hat{n}$ is immaterial, and the spots inside them do not count in computing n. So Fig. 2 has five regions, of types $\hat{5}, 5, \hat{3}, 3, 2$. What should be our move here?

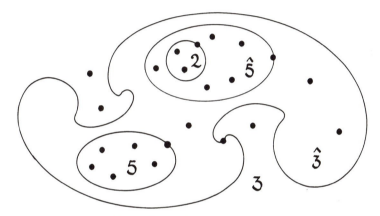

Figure 2. A Game of Contours.

The moves available from the general position are:

$$n \text{ or } \hat{n} \text{ to } a + \hat{b} \ (a > 0)$$

$$\hat{n} \text{ to } \hat{a} + \hat{b}$$

where in each case $a + b = n - 1$. So we can draw up a table of nim-values, as in Table 2.

n:	0	1	2	3	4	5	6	7	8	9	10	11	12	13	14	15	16	17	18	19	20	...
$\mathcal{G}(n)$:	0	1	0	1	0	3	2	0	5	2	0	1	4	3	2	0	5	2	3	1	...
$\mathcal{G}(\hat{n})$:	0	1	2	3	1	4	3	2	0	5	2	3	1	4	3	2	0	5	2	3	1	...

Table 2. Nim-values for Contours.

We see that for $n \geq 12$ the two nim-sequences coincide and have period 8. So the starting position with n spots is a $\mathcal{P}$-position in normal play only if n is 1, 3, 5, 11, or a multiple of 8. We have not found a complete misère analysis, but Table 3 gives the start of a genus analysis (Chapter 13).

n	0	1	2	3	4	5	6	7	8	9	10	11	12	13	14	15	16
genus of n		0	1	0	1	0	3	2	0	$5^{057}_{A_1}$	2^2_C	0^3	1^0	4^{146}	3^3	2^{20}	0^1
genus of $\hat{n}$	0	1	2	3	1	4^{146}_A	3	2	0^1_B	$5^{057}_{A_1}$	2^2_D	3^3	1^0	4^{146}	3^3	2^{20}	0^1

Table 3. The Genus of Contours Positions.

$A = 2_2321$, $B = A_2A_1321$. For every position with all numbers ≤ 10, the genus is correctly computed by pretending that $A + A = B = 0$, $C = D = 2$.

In the position of Fig. 2, the nim-values are 4, 0, 3, 0, 1, and so we must make a move changing nim-value 4 to 2. This can only be done by converting the region of type $\hat{5}$ into two of types $\hat{3}$ and $\hat{1}$, so we must draw a loop surrounding the inner contour of this region and just 1 or 3 more points, at least in normal play. It turns out that we have exactly the same good moves in misère play also.

Lucasta

This is an old game first described by Lucas, and since it does not seem previously to have had a proper name, we have named it for him. It is quite remarkable that we can give complete strategies for both normal and misère play from the starting position, although the general theory is very complicated. Our strategy for normal play is easily proved when once found, but the misère play strategy is very tricky indeed.

The move is to draw a curve having as endpoints two distinct spots. These may *not* be the two endpoints of a single previously drawn curve (though they *may* be linked together by a chain of curves through intermediate spots). No two curves may cross, and no spot may be an endpoint of more than two curves, so that the curves can only build up into chains or into closed loops which must go through three or more spots.

The loops separate the plane into connected regions, as in the previous games, but now the situation within one of these regions needs a triple (a, b, c) of three numbers to describe it adequately. Here a is the number of *atoms*, or isolated spots, b the number of *branches* joining two otherwise isolated spots, and c is the number of *chains* consisting of three or more spots joined by a sequence of edges. It turns out that the number of spots in a chain is immaterial, except that chains (3 or more spots) must be distinguished from atoms and branches (1 and 2 spots).

The possible moves are classified as follows:

aa: join two atoms to form a branch,
ab: join an atom to a branch, making a chain,
bb: connect two branches, forming a chain,
ac: lengthen a chain by adjoining some atom,
bc: extend a chain by attaching a branch,
$c!$:   *pull* a chain by joining its ends together.

Since pulling a chain divides a region into two, the result may depend on how it separates the remaining atoms, branches and chains from each other. We denote this by (for example) $c!(a^3)$ or $c!(ab)$, which mean that we separate 3 atoms or an atom and a branch into a region of their own. It is also possible to make a move cc: joining two chains to form a longer one: but the same effect could be achieved by *sharply* ($c!!$) pulling one of the chains, that is, with no separation.

A Child's Guide to Normal Lucasta

We were fortunate that the nim-values we computed for Lucasta suggested a pattern for the outcomes of all positions with at most one chain. This pattern is displayed in Table 4 in which the entry (a, b) is

P if $(a, b, 0)$ is a $\mathcal{P}$-position, so $(a, b, 1)$ is an $\mathcal{N}$-position
+ if $(a, b, 1)$ is a $\mathcal{P}$-position, so $(a, b, 0)$ is an $\mathcal{N}$-position
− if $(a, b, 0)$ and $(a, b, 1)$ are both $\mathcal{N}$-positions.

Notice that the columns repeat with period 4, after the first four, while the rows alternate.

$a =$	0	1	2	3	4	5	6	7	8	9	10	11	12	13	14	15
$b = 0$	P	P	−	+	P	P	P	−	P	P	P	−	P	P	P	−
1	P	P	−	−	−	P	−	−	−	P	−	−	−	P	−	−
2	P	P	−	+	P	P	P	−	P	P	P	−	P	P	P	−
3	P	P	−	−	−	P	−	−	−	P	−	−	−	P	−	−
4	P	P	−	+	P	P	P	−	P	P	P	−	P	P	P	−
5	P	P	−	−	−	P	−	−	−	P	−	−	−	P	−	−
6	P	P	−	+	P	P	P	−	P	P	P	−	P	P	P	−
7	P	P	−	−	−	P	−	−	−	P	−	−	−	P	−	−

Table 4. Lucasta with at Most One Chain.

A complete analysis may be difficult, though a machine attack will probably show that the nim-values have period 2 in b and c. However we can give a strategy which enables you to win all the positions you deserve to, when the number of chains is small. This strategy also proves that the pattern of Table 4 persists indefinitely. It uses the special $\mathcal{P}$-positions

$$(0, b, 0), \quad (1 + 4k, b, 0), \quad (3, 2m, 1), \quad (4 + 2k, 2m, 0) \quad \text{and} \quad (0, 2m, 2), \ b, k, m \geq 0.$$

It is almost always bad to leave chains in a position because they can be pulled in a number of different ways.

Our opponent can move from one of the special positions so as to leave two or more chains in only a few ways. If he joins two branches to form a chain we pull it smartly; if he joins a branch to a chain we join another. In either case the total effect is to remove two branches. The only other case is the ab move from $(3, 2m, 1)$ to $(2, 2m - 1, 2)$ after which we join the two atoms to get the position $(0, 2m, 2)$. Our responses to positions with at most one chain are given in Table 5. Observe that we have completely justified Table 4. The nim-values on which we based our strategy are shown in Table 6. The entry (a, b) gives the sequence of nim-values for $c = 0, 1, 2, \ldots$; the last pair of values always repeats indefinitely, so that 13145 abbreviates 131454545. . . . Each unprinted row in the first five columns has the same entries as that with b decreased by 2. We have shown that all $\mathcal{N}$-positions $(a, b, 0)$ except $(2, 2m + 2, 0)$ and $(6, 1, 0)$, have nim-value 1 and all $\mathcal{N}$-positions $(a, b, 1)$, except $(0, 2m, 1)$, $(1, 2m + 1, 1)$ and $(5, 0, 1)$ have nim-value at least 2.

	$a = 0,1,5,\ldots,1+4k$	$a = 2$	$a = 3$	$a = 4,6,\ldots,4+2k$	$a = 7,11,\ldots,7+4k$
$c = 0$	$\mathcal{P}$-positions; bad luck! Hope for a blunder	aa gives $(0, b+1, 0)$	aa gives $(1, b+1, 0)$	b even: bad luck! b odd: ab gives $(3, 2m, 1)$ if $k = 0$. aa gives $(2 + 2k, 2m + 2, 0)$ otherwise	aa gives $(5 + 4k, b+1, 0)$
$c = 1$	$c!!$ smartly gives $(0, b, 0)$ or $(1 + 4k, b, 0)$	$c!(a)$ round a solitary atom: $(1, 0, 0)+$ $(1, b, 0)$	b even: bad luck! b odd: bc gives $(3, 2m, 1)$	$c!$ separating atoms from branches, $(0, b, 0) + (4 + 2k, 0, 0)$	$c!$ separating all but one of the atoms from the branches: $(1, b, 0) + (6 + 4k, 0, 0)$

Table 5. How to Win at Lucasta.

$b = 0$	$a = 0$	1	2	3	4	5	6	7
0	01	023	13145	10201	0351732	01023245	0245713101	13169498
1	023	01	124567	13132	1464601	02518189	230645	154578Xx
2	01	023	2356745	10401	0258589	046262Tt	06798	1316XTFf
3	023	01	15478967	13132	1567Xx	020101tFf		
4	01	023	2376945	10401	0278549t98	046292TfTt		X = 10
								x = 11
5	023	01	15498X67	13132	15696x6xX	020101fF		T = 12
								t = 13
6	01	023	2376X45	10401	027854Tt98	0462X2tSs		F = 14
								f = 15
7	023	01	15498x67	13132	15696T6xSxX			S = 16
								s = 17
8	01	023	2376X45	10401	027854F89			
9	023	01	15498x67	13132	15696T6xSxX			

Table 6. Nim-values for Positions (a, b, c) in Lucasta.

The Misère Form of Lucasta

It is remarkable that we can still give a strategy for misére Lucasta from any starting position $(a, 0, 0)$. This is largely because the player who wins can do so without allowing the creation of too many chains, for of course positions with many chains are very difficult to analyze. For fairly small values of a, b, c we can of course compute the genus, as in Table 9, given later, which shows that the complete theory is very complicated. In fact, Table 9 was first used in constructing our other tables and figures, and it then suggested our general strategy. This strategy is described in Table 7, Fig. 3, Table 8 and the explanatory notes to these. In Table 7 the notation is as in Table 4, and the patterns continue.

a =	0	1	2	3	4	5	6	7	8	9	10	11	12	13	14	15	16	17	18	19
b = 0	+	−	P	P	−	P	−	P	−	P	−	−	P	P	−	−	−	P	−	−
1	−	+	P	−	P	−	P	−	P	P	−	+	P	P	P	−	P	P	P	−
2	+	P	−	P	−	P	−	−	P	P	−	−	−	P	−	−	−	P	−	−
3	P	−	P	−	P	P	−	+	P	P	P	−	P	P	P	−	P	P	P	−
4	−	P	−	−	P	P	−	−	−	P	−	−	−	P	−	−	−	P	−	−
5	P	P	−	+	P	P	P	−	P	P	P	−	P	P	P	−	P	P	P	−
6	P	P	−	−	−	P	−	−	−	P	−	−	−	P	−	−	−	P	−	−
7	P	P	−	+	P	P	P	−	P	P	P	−	P	P	P	−	P	P	P	−
8	P	P	−	−	−	P	−	−	−	P	−	−	−	P	−	−	−	P	−	−
9	P	P	−	+	P	P	P	−	P	P	P	−	P	P	P	−	P	P	P	−

Table 7. Outcomes of some Misère Lucasta Positions.

Table 7 gives the outcome of positions of form $(a, b, 0)$ or $(a, b, 1)$, and is the skeleton of our strategy. Most of the rest of our discussion is concerned only with the justification of the + entries. First we show how the remaining entries can be deduced from these. We use three principles.

(1) The entry (a, b) is P if and only if it is non-terminal and there is no entry of form

$$\begin{array}{llllll} & \text{P} & \text{in} & (a-2, b+1) & \text{the only} & aa & \text{to} & (a-2, b+1, 0) \\ & + & \text{in} & (a, b-2) & \text{moves from} & bb & \text{to} & (a, b-2, 1) \\ \text{or} & + & \text{in} & (a-1, b-1) & (a, b, 0) \text{ being} & ab & \text{to} & (a-1, b-1, 1). \end{array}$$

(2) The entry (a, b) cannot be + if there is any entry of form

P	in	$(a, b-2)$		$c!(bb)$	from $(a, b, 1)$	$(a, b-2, 0) + (0, 2, 0)$
P	in	$(a-1, b-1)$	because	$c!(ab)$	to each	$(a-1, b-1, 0) + (1, 1, 0)$
P or +	in	$(a-1, b)$	there are	$c!(a)$ or ac	of the	$(a-1, b, 0) + (1, 0, 0)$ or $(a-1, b, 1)$
P or +	in	$(a, b-1)$	moves	$c!(b)$ or bc	positions	$(a, b-1, 0) + (0, 1, 0)$ or $(a, b-1, 1)$

and the positions $(0, 2, 0)$, $(1, 1, 0)$, $(1, 0, 0)$, $(0, 1, 0)$ can be neglected, since they necessarily last for exactly 0 or 2 moves.

(3) The entry $(a, 0)$ cannot be + if there is a P entry in $(a-4, 1)$ or $(a-6, 0)$. (For from the position $(a, 0, 1)$ we can move to $(a-2, 0, 0) + (2, 0, 0)$, and whatever our opponent does to this, we can move to the sum $(a-4, 1, 0) + (0, 1, 0)$ on our next move, in which $(0, 1, 0)$ can be neglected. We can also move to $(a-6, 0, 0) + (6, 0, 0)$ and we shall show later that $(6, 0, 0)$ can be neglected, being equivalent to 0.)

The reader should now check that all the entries in Table 7 follow from the + entries using only these three principles, and the obvious fact that each entry *is* P or + or −, since $(a, b, 0)$ and $(a, b, 1)$ cannot both be P.

It is not such a routine matter to justify the + entries themselves, the main difficulty being that our opponent might try to create two or more chains, and we cannot allow this to persist, or the position will become too complicated for words (or pictures). The backbone of our strategy (supporting the skeleton of Table 7) is shown in Fig. 3, which illustrates winning strategies for the second player from each of the positions

$$(0, 0, 1), (1, 1, 1) = (0, 2, 1), (3, 5, 1), (3, 7, 1), (3, 9, 1), \ldots .$$

which are all but two of the $\mathcal{P}$-positions corresponding to the + entries in Table 7. We write $(1, 1, 1) = (0, 2, 1)$ because a single atom has exactly the same effect on the game as another branch. For the same reason, we have systematically replaced any position $(1, b, c)$ which should appear in Fig. 3 by the equivalent position $(0, b + 1, c)$.

Some further remarks need to be made about Fig. 3. The positions surrounded by double boxes represent $\mathcal{P}$-positions which will be dealt with shortly. All other $\mathcal{P}$-positions are surrounded by single boxes, and all their options also appear in the figure. Every unboxed position on the diagram represents an $\mathcal{N}$-position, for which a $\mathcal{P}$-option is always given. The symbol $abcD$ denotes the sum of the position (a, b, c) with another position (like $(0, 0, 1)$) which necessarily lasts an odd number of moves (usually *one* move), however played, while $abcE$ denotes the sum of (a, b, c) with a position (like $(0, 2, 0)$ or $(1, 1, 0)$) which necessarily lasts an *even* number of moves (usually *two*). In any later analysis, we have always supposed that these odd and even numbers were *one* and *zero*, respectively. Finally, $^*abc$ denotes the sum of any two positions (x, y, z) and $(a - x, b - y, c - z)$. To continue the figure downwards increase b by 2.

The two $\mathcal{P}$-positions $(7, 3, 1)$ and $(11, 1, 1)$ corresponding to the only + entries in Table 7 not yet verified, are discussed in Table 8.

It remains to discuss the double-boxed positions of Fig. 3.

Theorem. The sum of any number of positions of the form $(0, b, 0)$ together with a game which necessarily lasts for exactly n moves, is a $\mathcal{P}$-position if and only if:

  either n is odd, and all the numbers b are 0, 1, 2, or 4

  or n is even, and at least one of the numbers b is *not* 0, 1, 2, or 4.

Proof. The positions $(0, 0, 0)$ and $(0, 1, 0)$ are ended, while $(0, 2, 0)$ lasts exactly two moves, so all of these positions can be neglected. In fact positions $(0, 4, 0)$ can be neglected as well, since we can always arrange that they last an even number of moves. The only line of play from $(0, 4, 0)$ lasting an odd number of moves is

$$(0, 4, 0) \text{ to } (0, 2, 1) \text{ to } (0, 1, 1) \text{ to } (0, 1, 0).$$

We need never make the move from $(0, 2, 1)$ to $(0, 1, 1)$, and if our opponent does so, we can immediately reply with a move from $(0, 1, 1)$ to $(0, 0, 1)$ which makes the game last an extra move.

Neglecting $(0, 4, 0)$ and positions which always last an even number of moves, the only real assertion is that a sum of positions $(0, b, 0)$, with each b either $= 3$ or ≥ 5, is a $\mathcal{P}$-position. The only move from $(0, b, 0)$ is to $(0, b - 2, 1)$ from which we can move to any position $(0, x, 0) + (0, y, 0)$ with $x + y = b - 2$. However our opponent moves, we can use this to restore the position to another one covered by the theorem, *unless* it is just the single position $(0, 3, 0)$, from which our opponent can only move to $(0, 1, 1)$, and we then move to $(0, 0, 1)$, leaving him to make the last (losing) move.

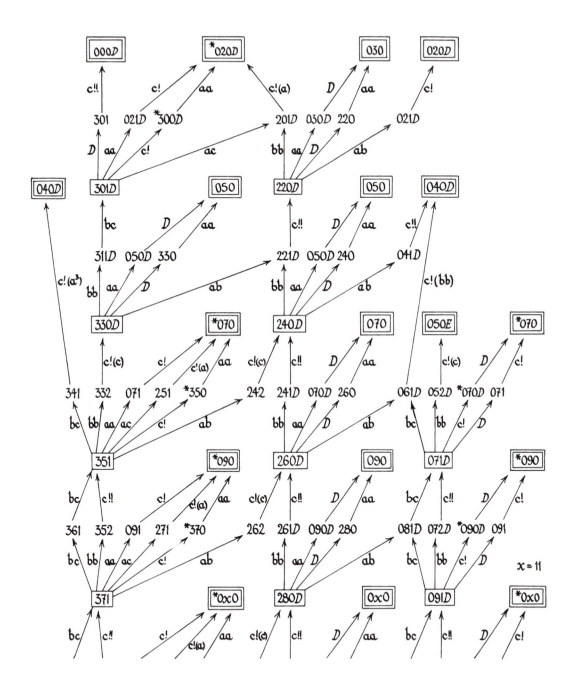

Figure 3. Strategy for Misère Lucasta.

The Positions (7, 3, 1) and (11, 1, 1)

In Table 8, to each option of either of these positions, we give a response, which is in every case expressed as the sum of a $\mathcal{P}$-position from Table 7 and some position which can be verified to be equivalent to 0 from Fig. 4. In fact the $+$ entries in Table 7 corresponding to $(7, 3, 1)$ and $(11, 1, 1)$ are not needed to verify any other entry, and so are not used in our strategy from any initial position, so that Table 8 is not really necessary for the strategy.

The options of (7, 3, 1)	have good replies	The options of (11, 1, 1)	have good replies
aa $(5, 4, 1)$	$(5, 2, 0) + (0, 2, 0)$	$(9, 2, 1)$	$(9, 0, 0) + (0, 2, 0)$
ab $(6, 2, 2)$	$(3, 2, 0) + (3, 0, 1)$	$(10, 0, 2)$	$(7, 0, 0) + (3, 0, 1)$
bb $(7, 1, 2)$	$(1, 1, 1) + (6, 0, 0)$		
ac $(6, 3, 1)$	$(5, 3, 0) + (1, 0, 0)$	$(10, 1, 1)$	$(9, 1, 0) + (1, 0, 0)$
bc $(7, 2, 1)$	$(7, 0, 0) + (0, 2, 0)$	$(11, 0, 1)$	$(7, 0, 0) + (4, 0, 0)$

The other options are all sums of two positions $(a, b, 0)$, and in every case we move in the region which has just 2, 3, 6, 7, 10, or 11 atoms, and join two of these atoms together.

Table 8. $(7, 3, 1)$ and $(11, 1, 1)$ are $\mathcal{P}$-positions.

It is interesting to note that the positions

$$000, 010 = 100, 020 = 110, 040 = 130,$$
$$400, 420, 510, 600, 800,$$
$$301, 022 = 112, 002, 004, 006, \ldots$$

are equivalent to 0 in the misère sense. (This remark can be useful in play from more complicated positions than those which need arise if our strategy is followed.) To prove that a position is misère-equivalent to 0 it is necessary and sufficient to show, first, that it is an $\mathcal{N}$-position and second, that each of its options has itself an option misère-equivalent to 0. This is done for the above positions in Fig. 4, in which the subscript

P denotes a $\mathcal{P}$-position
N denotes an $\mathcal{N}$-position *not* misère-equivalent to 0
O denotes an $\mathcal{N}$-position misère-equivalent to 0.

In a strategically fought game of Misère Lucasta, we find three phases. In the first phase, both players join pairs of atoms together to form branches. If either player dares to form a chain, his opponent can certainly win by closing the chain around some small number of atoms and branches (which can be neglected), and converting the rest of the position to a $\mathcal{P}$-position. When the number of atoms is reduced to just above three, the winner is the player able to convert the position to $(3, 2n + 1, 1)$, and the game enters its second phase, in which play follows the lines of Fig. 3. The third phase is reached when the position becomes a sum of positions $(0, b, 0)$ in which only branches (with isolated atoms) remain, together possibly with some rather trivial game. From then on, the winner always restores the position to a similar form, except that near the end of the game he is careful to restore the position $(0, 3, 0)$ to a single chain $(0, 0, 1)$.

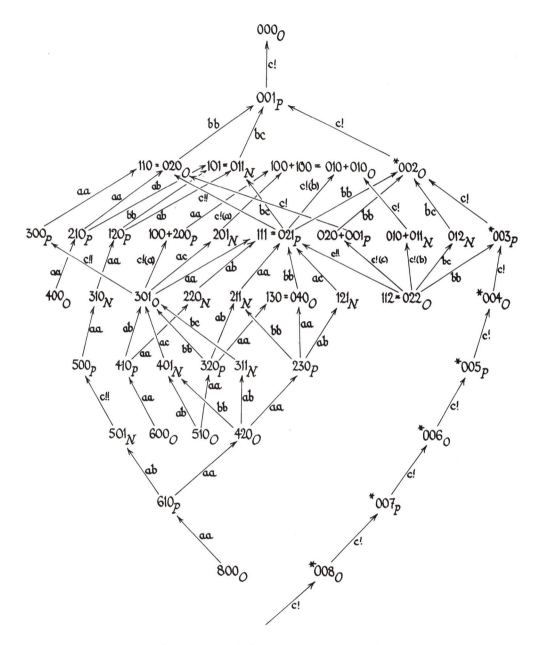

Figure 4. Proof of Misère-equivalence to Zero.

Table 9. The Genus of Lucasta Positions.

$a = 0$

$b \backslash c$	0	1	2	3	4	5	6	7	8
0	0	1	0	1	0	1	0	1	0
1	0	2	3	2	3	2	3	2	3
2	0	1	0	a	a_1	a	a_1	a	a_1
3	2_+	2	3	b	c	d	d_1	d	d_1
4	0	e	f	g	h	i	j	k	k_1
5	2_+	l	3^{04}						
6	e_+	1^3	0^{52}						
7	l_+	2^1							
8	0^0	1^2							
9	0^0								
10	0^0								

$a = 1$

$b \backslash c$	0	1	2	3	4	5	6	7	8
0	0	2	3	2	3	2	3	2	3
1	0	1	0	a	a_1	a	a_1	a	a_1
2	2_+	2	3	b	c	d	d_1	d	d_1
3	0	e	f	g	h	i	j	k	k_1
4	2_+	l	3^{04}						
5	e_+	1^3	0^{52}						
6	l_+	2^1							
7	0^0	1^2							
8	0^0								
9	0^0								

$a = 2$

$b \backslash c$	0	1	2	3	4	5	6
0	1	3	1	p	p_1	p	p_1
1	1	2	4	q	r	s	s_1
2	t	u		v	6^{686}		
3	1	w		4^{04}			
4	K	3^3		5^{16}			
5	1^1	5^4					
6	2^1	3^{203}					
7	1^2						
8	2^1						

genus	name	structure
1^{4313}	a	2_2320
2^{2020}	b	$a2_+30$
3^{0431}	c	$ba_1a2_{+1}2$
2^{1520}	d	$cb_1a_2a_ia2_+3$
1^{3131}	e	2_+20
0^{1202}	f	$ea2_{+2}321$
1^{4313}	g	$fe_1ba_12_{+3}2_230$
0^{5202}	h	$gf_1ecba2_{+2}3_21$
1^{4313}	i	$hg_1fe_1dcb_2a_12_{+3}2_20$
0^{5202}	j	$ih_1gf_1ed_1dc_2b_3a2_{+2}3_21$
1^{4313}	k	$ji_1hg_1fe_1d_2d_1dc_3b_2a_aa_12_{+3}2_20$
0^{0202}	a_a	$a_{22}a_3a_2a$
2^{2020}	l	$e2_+30$
4^{1464}	p	2_2321
5^{5757}	q	$pa3_243210$
6^{6846}	r	$qp_1pa_1a2_25320$
7^{7957}	s	$rq_1p_2p_1pa_2a_1a3_24321$
2^{1420}	t	2_+31
3^{3131}	u	$t2_+210$
5^{5757}	v	$ut_1pa2_{+2}43210$
5^{2057}	w	$ute2_{+1}2_+4310$

genus	name	structure
0^{3131}	A	$pa3_221$
1^{2020}	B	$Ap_1pa_12_230$
0^{3131}	C	$BA_1p_2p_1pa3_221$
1^{5313}	D	$p2_+4320$
3^{6464}	E	$DAqp_1b2_{+1}2_2421$
1^{1313}	F	$2_+3 = 2_{+1}$
1^{2020}	H	F_+30
2^{0313}	I	$H0$
1^{1313}	J	$u2_+3$
2^{1313}	K	$ue2_+$
0^{0202}		$2_+, e_+, l_+, F_+$

		$a = 3$						$a = 4$		$a = 5$		$a = 6$		7	8	9
$c =$	0	1	2	3	4	5	6	0	1	0	1	0	1	0	0	0
$b = 0$	1	0	2	A	B	C	C_1	0	3	F_+	H	0	2^3	1	0	0^0
1	F	3	D	E				1	4^4	0		I		1^1		
2	1							0		0^0						
3	J															

Note: If $c \geqslant 2b + 2a$, then (a, b, c) has value x_1, where $(a, b, c-1)$ has value x.

To save space in Table 9 the abbreviating conventions are *not* the same as those of Chapter 13 (see Volume 2 of Winning Ways). In the table

$$g^{a \ldots x} \quad \text{means} \quad g^{a \ldots xyxy\ldots} \quad \text{where} \quad y = x \overset{*}{+} 2$$

and no assertions about tameness or restiveness are intended. In the notes opposite, the genus is given to four superscripts even when the period starts earlier, and now the last two superscripts repeat indefinitely.

Cabbages; or Bugs, Caterpillars and Cocoons

If we modify Lucasta by allowing the move which completes a closed loop passing through only two spots and consisting of two curves joining them, we get a simpler game. Here, we call the isolated spots *bugs*, chains of two or more spots *caterpillars*, and closed loops *cocoons*. The cocoons separate the plane into regions, so that the general position is a sum of positions (b, c), where these numbers specify the numbers of bugs and caterpillars per region.

It turns out that the position (b, c) in this game behaves just like the position $(0, b, c)$ in Lucasta, so we have the analysis already. (Using our nim-value table for Lucasta we can in fact analyze arbitrary positions in normal play.) In particular, we have:

The initial position $(n, 0)$ is a $\mathcal{P}$-position in normal play for all n,
and   in misère play for all n except 0, 1, 2, 4.

Jocasta

We obtain an even simpler game by allowing in addition the move which joins an isolated spot to itself to form a closed loop passing only through that spot.

Sprouts

This game (introduced by M. S. Paterson and J. H. Conway some time ago) has a novel feature which complicates the analysis to such an extent that the normal outcome of the 7-spot game remained unknown until 1999. Even the 2-spot game is remarkably complicated.

The move in Sprouts is to join two spots, or a single spot to itself (Fig. 5) by a curve which does not meet any previously drawn curve or spot. But when this curve is drawn, a new spot must be placed upon it. No spot may have more than three parts of curves ending at it.

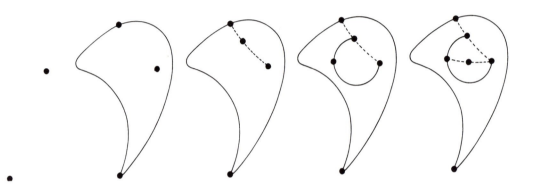

Figure 5. A Short Game of Sprouts.

A typical game is shown in Fig. 5, with the second player's moves drawn as dotted lines. Since the new spots can still be used in later moves, a Sprouts game will last longer than a Cabbages game from the same initial position, and it is perhaps not even obvious that it need ever end. But there is a simple argument which shows that in fact a Sprouts game starting from n spots can last at most $3n - 1$ moves. We take the 3-spot game as an example. Each spot has potentially 3 ends of curves available to it, which we shall call its three *lives*, so initially the 3-spot game has 9 lives. But each move takes one life away from the two spots it joins (or two lives away from a spot joined to itself), and adds a new spot which has just one life. Therefore each move reduces the total number of lives by one. Since the very last spot to be created is still alive at the end of the game, the total number of moves is at most $9 - 1 = 8$. But Fig. 6 shows just how complicated even the 2-spot game really is.

One of the most interesting theorems about Sprouts (due to D. Mollison and J.H. Conway) is the Fundamental Theorem of Zeroth Order Moribundity (FTOZOM). We shall not prove it here, but will at least state it. The FTOZOM asserts that the n-spot Sprouts game must last at least $2n$ moves, and that if it lasts exactly this amount, the final configuration is made up of the insects shown in Fig. 7.

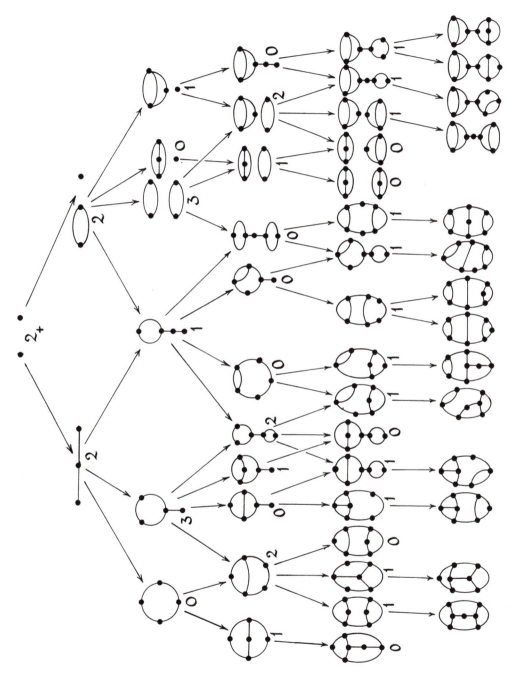

Figure 6. Two-spot Sprouts, with Reduced Forms.

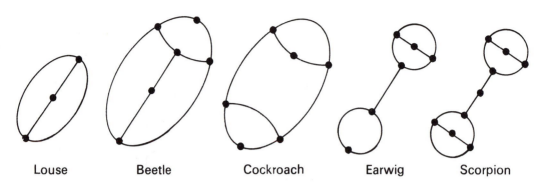

Figure 7. The Five Fundamental Insects.

To be more precise, the final configuration must consist of just one of these insects (which might perhaps be turned inside out in some way) infected by an arbitrarily large number of lice (some of which might infect others). One of the possible configurations is shown as Fig. 8—it consists of an inside-out scorpion inside an inside-out louse, liberally infested with other lice!

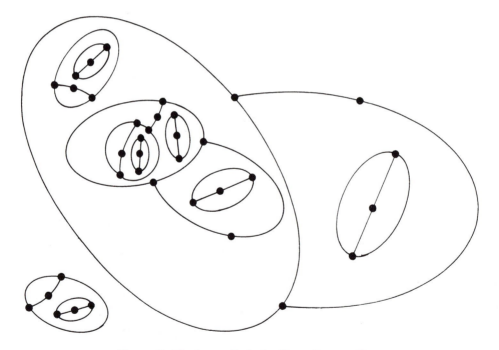

Figure 8. The Lousy End of a Short Sprouts Game.

How should we play if we wish to win a Sprouts game? It is clear that whether the play is normal or misère, the outcome only depends on whether the total number of moves in the game is odd or even, so in some sense winning is controlling the number of moves. Now the

3-spot game necessarily lasts for 6, 7, or 8 moves, and it is very difficult to make it last 8 moves, so that really the fight is between 6 and 7 moves. Apparently the same thing happens in larger games—essentially one player tries to make the game last m moves, while the other tries to drag it out to $m + 1$, all other numbers being very unlikely.

To see how to control the number of moves, we examine the situation at the end of the game, which we suppose to have started with n spots and lasted for m moves. The final number of spots is $n + m$, and the total life at the end of the game is $l = 3n - m$, since we started with $3n$ lives, and subtracted one per move. Each of the live spots at the end of the game has two dead spots as its two nearest neighbors, and the remaining dead spots are called *Pharisees*. (The concept of neighbor is quite subtle—in Fig. 9 we show the two different ways in which two dead spots can be neighbors of a live one.)

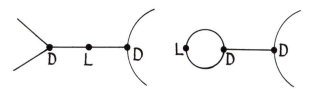

Figure 9. Two Live Spots (L) and Their Dead Neighbors (D).

Now no dead spot can be a neighbor of two different live spots , for otherwise we could join these two spots and continue the game. So the number ϕ of Pharisees is given by the equation

$$\phi = (n + m) - (l + 2l) = (n + m) - 3(3n - m) = 4m - 8n$$

and we have the *Moribundity Equation*:

$$m = 2n + \frac{1}{4}\phi.$$

From this equation we can deduce several things:

(i) The number of moves is at least $2n$.

(ii) The number of Pharisees is a multiple of 4.

(iii) If at any time in the game we can ensure that the final position has at least P Pharisees, then the game will last at least $2n + \frac{1}{4}P$ moves.

There is a corresponding result to (iii) in the opposite direction:

(iv) If at any time in the game we can ensure that the final position has at least l live spots, then the game will last at most $3n - l$ moves.

So, according to our previous ideas, one player will try to lengthen the game by producing Pharisees, while his opponent tries to shorten it by producing spots which must remain alive.

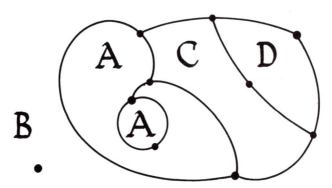

Figure 10. A Sprouts Position with One Pharisee.

There is a useful way to estimate the number of live spots there will be at the end of the game. *If any region defined by curves of the game has a live spot strictly inside, then there will be a live spot inside that region at all later times.* So in Fig. 10 we can regard, if we like, the plane as divided into four regions A, B, C, D, and the regions A and B each have live spots strictly inside. Any move made in either of these regions creates a new live spot, and so each of A and B will contain a live spot at the end of the game. We cannot say the same of C and D, whose only live spots lie on their borders, but if we regard C and D together as forming a single region, then this new region has just one spot strictly inside. So we can see that the game will have at least 3 live spots in its final position. It also has presently one Pharisee P, and so (since it developed from an initial position with $n = 4$ spots) we can see that it will last *at most* $3n - 3 = 9$ moves, and *at least* $2n + \frac{1}{4} = 8\frac{1}{4}$ moves. Since it is difficult to see how the game could last for *exactly* $8\frac{1}{4}$ moves, we conclude that the total length of the game will be 9 moves, however it is played from now on! (Actually, 6 moves have already been made, so just 3 more moves are to follow.) Accordingly, this is either a normal play game about to be won by the first player, or a misère play one being won by the second player.

Using computers, Applegate, Jacobson and Sleator have extended our results considerably. They make the rather slender conjecture that Sprouts is a first player win just if the initial number of dots is $\equiv 3$, 4 or 5 (mod 6). They have verified this for $n < 12$. The corresponding misère Sprouts conjecture is that the first player has a win just if $n \equiv 0$ or 1 (mod 5). This they verified for $n < 10$.

no. of spots:	0	1	2	3	4	5	6	7	8	9	10	11
normal play:	$0P$	$2P$	$4P$	$7N$	$9N$	$11N$	$14P$	P	P	N	N	N
misère play:	$0N$	$2N$	$5P$	$7P$	$9P$	N	N	P	P	P		

Table 10. Outcomes of the Smallest Sprouts Games.

The fact that 6-spot normal Sprouts is a $\mathcal{P}$-position was first proved (to win a bet) by Denis Mollison, whose analysis of the game ran to 47 pages! Using the ideas above, we can shorten this considerably, but 5-spot Sprouts with misère play still seems to need a computer.

Brussels Sprouts

Here is another game, which should be more interesting than Sprouts. We start with a number of crosses, instead of spots. The move is to continue one arm of a cross by some curve which ends at another arm of the same or a different cross, and then to add a new cross-bar at some point along this curve. A 2-cross game of Brussels Sprouts is shown as Fig. 11. After playing a few games of Brussels Sprouts, the skillful reader will be able to suggest a good starting strategy.

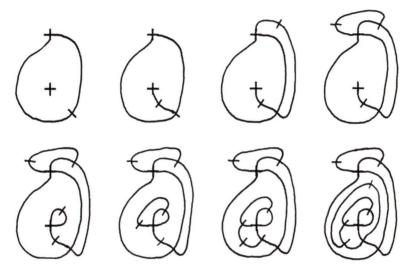

Figure 11. A 2-cross Game of Brussels Sprouts.

Stars-and-Stripes

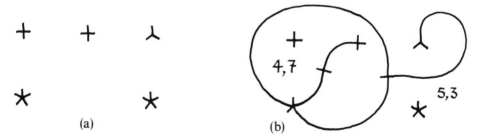

Figure 12. A Game of Stars-and-Stripes.

Suppose we make addition of the cross-bar optional in Brussels Sprouts. It is natural at the same time to allow "stars" with any number of arms instead of just crosses with exactly 4 arms, and to call the cross-bar a *stripe*. An initial position (5, 5, 4, 4, 3) is shown in Fig. 12(a), along with a position 3 moves later (Fig. 12(b)). In the analysis, the game becomes a disjunctive sum of regions, and we can pretend that each region contains only stars. In general, a connected portion of the picture which has just n arms sticking into the region concerned counts as an n-arm star inside the region. (Even the boundary of the region counts as a star.) In Fig. 12(b) we have therefore labelled each region with numbers showing the sizes of the stars in that region.

The one-star game is isomorphic to the octal game **4·07**, since a move with cross-bar essentially splits an n-arm star into two stars of sizes a and b, with $a + b = n$, a, $b \neq 0$, and the move without crossbar splits it into stars of sizes a and b with $a + b = n - 2$. The nim-values for this game (Chapter 4, Tables 7(a), 6(b)) are $0.\dot{0}12\dot{3}$ and the genus appears in Table 11.

n	0	1	2	3	4	5	6	7	8	9	10	11
genus of n	0	0	1	2	3	0	1_a^{431}	2	3_b^{31}	$0_{a_1}^{520}$	1_c^{431}	2_d^{0420}

$$a = 2_2 320 \quad b = a_1 a 2_2 20 \quad c = b_1 b a_3 a_1 2_2 20 \quad d = c b_2 b a_2 a_1 a 3_2 3$$

Table 11. The Genus of Stars-and-Stripes Positions.

Bushenhack

Bushenhack is another pencil and paper game. It's played with a number of rooted trees, but now when you chop an edge, all edges connecting it to the ground disappear, leaving a number of floating bits of tree to be rerooted as in Fig. 13. Its theory involves yet another property of Nim.

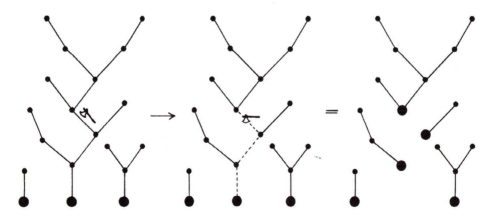

Figure 13. A Bushenhack Move.

Genetic Codes for Nim

If you tell me that you're in a Nim-position of some nim-value (e.g. 9) and can move to positions having exactly so many (e.g. 13) other nim-values, I can tell you exactly what (in this case 0, 1, 2, 3, 4, 5, 6, 7, 8, 12, 13, 14, 15) those values are!

To see why, we enlarge upon a notation from the Extras to Chapter 7, in which the single Nim-heaps are

$$0_{\{\}}, 1_{\{1\}}, 2_{\{2,3\}}, 3_{\{1,2,3\}}, 4_{\{4,5,6,7\}}, 5_{\{1,4,5,6,7\}}, \ldots,$$

and in general $n_{[n]}$ where $[n]$, the *variation set*, is the set of changes in nim-value that are possible in one move. The variation set $[n]$ for an arbitrary Nim-heap consists of all numbers whose leftmost binary digit 1 is present in the binary expansion of n, and so it can be found as the union of the appropriate selection of

$$[1] = \{1\}, \quad [2] = \{2, 3\}, \quad [4] = \{4, 5, 6, 7\}, \quad [8] = \{8, 9, \ldots, 15\}, \quad \ldots$$

E.g., since $13 = 1 + 4 + 8$, $[13] = \{1, 4, 5, 6, 7, 8, 9, \ldots, 15\}$.

We'll say that a position has **genetic code** A if it has the same variation set as the Nim-heap of size A. Arbitrary Nim-positions have genetic codes, because when you *add* positions you *unite* their variation sets; e.g., $5 + 12$ has genetic code 13, because

$$5 + 12 = 5_{\{1,4,5,6,7\}} + 12_{\{4,5,6,7,8,9,\ldots,15\}} = 9_{\{1,4,5,6,7,8,9,\ldots,15\}} = 9_{[13]},$$

and the options of $9_{[13]}$ can be found by nim-adding 9 to the members of the variation set $[13]$.

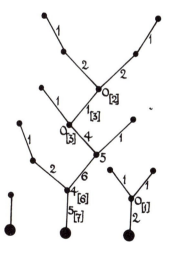

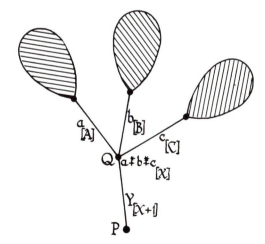

Figure 14. What's the Winning Move? (See the Extras.)

Figure 15. Calculating the Value and the Code.

Bushenhack Positions Have Genetic Codes!

In Fig. 14 the symbol $a_{[A]}$ against any edge gives the value and genetic code for the subtree whose trunk is that edge, while at every node where there are several branches we've given this information for the sum of the corresponding subtrees. (An isolated digit a means $a_{[a]}$.) The numbers are calculated as in Fig. 15 where X is the number whose binary expansion has a 1 wherever there is a 1 in *any* of A, B, C, so that

$$[X] = [A] \cup [B] \cup [C]$$

and Y is the smallest number greater than $a \overset{*}{+} b \overset{*}{+} c$ that is divisible by exactly that power of 2 which divides $X + 1$.

Suppose that X has binary expansion

$$\dots\, ?\ ?\ ?\ 0\ 1\ 1\ 1\ \dots\ 1 \quad \text{(ending in } k \text{ 1's)}.$$

Then that of $a \overset{*}{+} b \overset{*}{+} c$ will have the form

$$\dots\ p\ q\ r\ 0\ t\ u\ v\ \dots\ z$$

We already know why these numbers are the code and value for the sum of the three subtrees above Q in Fig. 15. So we need only show that

$$X + 1 \quad = \dots\ ?\ ?\ ?\ 1\ 0\ 0\ 0\ \dots\ 0$$

and

$$Y \quad = \dots\ p\ q\ r\ 1\ 0\ 0\ 0\ \dots\ 0$$

are the code and value for the subtree with trunk PQ. The options for this tree have nim-values

$$a \overset{*}{+} b \overset{*}{+} c \quad (\text{chop } PQ) \quad \text{and} \quad a \overset{*}{+} b \overset{*}{+} c \overset{*}{+} \Delta$$

for any number Δ whose first digit 1 is present in X. In particular, we can move to all nim-values that are $\leq a \overset{*}{+} b \overset{*}{+} c$ or differ from it only in the last k places. So we can move to all numbers smaller than Y, which differs from $a \overset{*}{+} b \overset{*}{+} c$ in the $(k + 1)$st place from the right, corresponding to the rightmost zero in X. The nim-values of the options are exactly those binary numbers whose leftmost difference from Y corresponds to a digit 1 in $X + 1$, which is therefore the genetic code.

Von Neumann Hackenbush

When played on trees, von Neumann Hackenbush is an exactly equivalent game in which the move is to delete a *node* together with all nodes on the path connecting it to the ground and all edges meeting these. To convert to Bushenhack just add a new trunk to every tree. Von Neumann proved, by a strategy-stealing argument, that a single tree was always an $\mathcal{N}$-position, and Úlehla gave an explicit strategy for trees, which prompted our own discussion.

Bushenhack is really just the theory of $A + B$ and $A{:}*$, whereas ordinary Hackenbush is concerned with $A + B$ and $*{:}B$. The most general version of von Neumann Hackenbush is played on any directed graph (remove a node and all nodes it points to). Its analysis involves the properties of $A + B$ and $A{:}B$ for arbitrary variation sets (Extras to Chapter 7).

Extras

The Joke in Jocasta

The Joke in Jocasta is that the n-spot game always lasts for n moves, because each spot has two lives and each move uses two. The game is therefore just another form of She-Loves-Me, She-Loves-Me-Not.

The Worm in Brussels Sprouts

The Worm in Brussels Sprouts is similar but more subtle. The n-cross game always lasts for just $5n - 2$ moves, but Brussels Sprouts is definitely more interesting on surfaces of higher genus, e.g. the torus.

Bushenhack

The winning move in Fig. 14 is that shown in Fig. 13.

References and Further Reading

Hugo D'Alarcao and Thomas E. Moore, Euler's formula and a game of Conway, *J. Recreational Math.* **9**(1977) 249–251; Zbl. 355.05021.

Piers Anthony, *Macroscope*, Avon, 1972.

David Applegate, Guy Jacobson and Daniel Sleator, Computer analysis of Sprouts, pp.199–201 in E. R. Berlekamp and Tom Rodgers, editors, *The Mathemagician and Pied Puzzler*, A K Peters, Natick MA, 1999.

Mark Copper, Graph theory and the game of Sprouts, *Amer. Math. Monthly*, **100** (1993) 478–482; MR 94c:90137.

Susan K. Eddins, Networks and the game of Sprouts, *NCTM Student Math. Notes*, May/June 1998.

Martin Gardner, Mathematical Games: Of sprouts and Brussels sprouts; games with a topological flavor, *Sci. Amer.* **217** #1 (July 1967) 112–115.

Martin Gardner, *Mathematical Carnival*, Alfred A. Knopf, New York 1975, Chapter 1.

T. K. Lam, Connected Sprouts, *Amer. Math. Monthly*, **104** (1997) 116–119; MR 98e:90251.

E. Lucas, *Récréations Mathématiques*, Gauthiers-Villars, 1882–94; Blanchard, Paris, 1960.

Gordon Pritchett, The game of Sprouts, *Two-Year Coll. Math. J.* **7** #4 (Dec. 1976) 21–25.

J. M. S. Simões-Pereira and Isabel Maria S.N. Zuzarte, Some remarks on a game with graphs, *J. Recreational Math.* **6**(1973) 54–60; Zbl. 339.05129.

J. Úlehla, A complete analysis of von Neumann's Hackendot, *Internal. J. Game Theory*, **9**(1980) 107–115.

-18-

The Emperor and His Money

Figure 1. The Emperor's Declaration.

"...Emperor Nu took power by overthrowing the divisive My-Nus dynasty. The Nu régime introduced many positive reforms, and in particular abolished the old (An-Tsient) irrational currency, which had his predecessor's head on it, and introduced the Nu system. The masters of the Imperial Mint, Hi and Lo, were alternately to decide the value of each new denomination, and after each decision, sufficiently many coins of this value were to be struck. All went well until Hi ordered the striking of a coin of value one, so throwing the Workers of the Mint into unemployment. They rose in a body, and threw the unfortunate Hi from the tower at the quiet end of the capital, which has been known as the Hi Tower ever since."

My-Nus—Some Divisive Times.

Sylver Coinage

Had Hi and Lo read this book, they would have realized they were only playing a game, the game of **Sylver Coinage**. In this the players alternately name different numbers, but are not allowed to name *any* number that is a sum of previously named ones. So, if 3 and 5 have been named, for example, neither of the players is allowed to play any of the numbers

$$3, \quad 5, \quad 6 = 3+3, \quad 8 = 3+5, \quad 9 = 3+3+3, \quad 10 = 5+5, \quad 11 = 3+3+5, \quad \dots$$

When will this game end? If neither player has played 1, 1 will still be playable. But, of course, as soon as 1 has been played, every number

$$1, \quad 2 = 1+1, \quad 3 = 1+1+1, \quad 4 = 1+1+1+1, \quad 5 = 1+1+1+1+1, \quad \dots$$

is illegal, and so the game ends. Because the player who names 1 is declared the *loser*, Sylver Coinage is a *misère* game. (Skillful players won't spend much time on the normal play version!).

We had better point out that because the old currency had been rather irrational (with coins of value $\sqrt{2}$, e and π) the Emperor declared that there was to be a new monetary unit, the **You-Nit**, and the value of each coin was to be an integral number of You-Nits. (You can see the Emperor making this declaration in Fig. 1!).

And recalling how people were nonplussed by the great financial scandal of the My-Nus dynasty when they had to take away Teh Kah-Weh for issuing currency of negative value, Emperor Nu decided that each coin's value must be a *positive* number of You-Nits.

How Long Will It Last?

It might take quite a long time. To see that it can last for a thousand moves, we need only consider the game
$$1000, 999, 998, \dots, 4, 3, 2, 1.$$

And of course a thousand can be replaced by any other number, so that the game is **unbounded**. Many other games have this property, for example Green Hackenbush (Chapter 2) played with an infinite snake, but are *boundedly* unbounded because after some fixed number of moves the end will be in sight. Thus after the first move in the Hackenbush game only a finite amount of snake is left.

But Sylver Coinage is not like that! No matter what number you choose. Hi and Lo can find a way to play that number of moves so that what's left of the game will still be unbounded. Their first thousand moves might be
$$2^{1000}, \ 2^{999}, \ 2^{998}, \dots, 2^4, 2^3, 2^2, 2^1$$

and the rest of the game can still last as long as you like:
$$1000001, 999999, 999997, \dots, 7, 5, 3, 1.$$

In other words Sylver Coinage is *unboundedly unbounded*. And this isn't all. It's *unboundedly unboundedly unbounded* and unboundedly like that, and so (unboundedly) on!

Nevertheless, it can't go on for ever; in the language of Chapter 11 it's an *ender*. It is because the little theorem which proves this is due to the famous mathematician J.J. Sylvester that we have called the game *Sylver* Coinage.

For, at any time after the first move, let g be the greatest common divisor (g.c.d.) of the moves made. Then it's not hard to see that only finitely many multiples of g are *not* expressible as sums of numbers already played. So after at most this known number of moves the g.c.d. must be reduced. Eventually we must arrive at a position with $g = 1$ and can bound the number of moves yet to be made. So although we may not be able to bound the game after any given number of moves, we *can* bound the number of moves it will take to reduce the g.c.d.

Some Openings Are Bad

The proof we gave in the Extras to Chapter 2 shows that from any position in Sylver Coinage there *is* a winning strategy for one of the two players *but* because of the infinite nature of the game we cannot work through all positions and guarantee to find winning strategies when they exist. In fact we do not know of (and there may not exist) any way of working out in a finite time who wins from an arbitrarily given position. But we do know the answers for some easy positions.

If at any time you name 1, you lose by definition.

If you name 2, my reply will be 3 if it's still available, and then all larger numbers

$$4 = 2+2, \qquad 5 = 2+3, \qquad 6 = 2+2+2, \qquad 7 = 2+2+3, \qquad 8 = 2+2+2+2, \quad \ldots$$

are excluded and you will be forced to name 1.

If you name 3, then for the same reason, 2 is a good reply.

So whoever first names any of 1, 2 and 3 will lose. In particular the first three numbers are bad opening moves. What will you reply if I open with 4? Maybe 5? If so the g.c.d. becomes 1 and there will be only finitely many numbers left. We can find out which by arranging the numbers as in Fig. 2. The circled numbers are excluded because they're multiples of 5 and these exclude the lower numbers by adding 4's. So only 1, 2, 3, 6, 7, 11 remain.

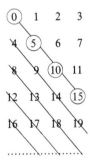

Figure 2. What's Left after $\{4, 5\}$.

I won't take 1, 2 or 3. If *I* say 6 or 7, you'll say the other, since these dismiss 11 and leave only 1, 2, 3 for me. So I'll say 11 and make *you* say 6 or 7 instead.

$$\boxed{\{4, 5, 11\} \text{ is a } \mathcal{P}\text{-position.}}$$

Here's what happens after 4 and 6.

$$
\begin{array}{cccc}
⓪ & 1 & 2 & 3 \\
\cancel{4} & 5 & ⑥ & 7 \\
\cancel{8} & 9 & \cancel{10} & 11 \\
\cancel{12} & 13 & \cancel{14} & 15 \\
\cancel{16} & 17 & \cancel{18} & 19
\end{array}
$$

....

Since 5 and 7 exclude all large numbers, they kill each other. Similarly for 9 and 11, and for 13 and 15, and so on.

> After {4,6} the pairs
> $(2, 3), (5, 7), (9, 11), \ldots, (4k + 1, 4k + 3)$,
> for $k \geq 1$, are mates.

So if you open with 4, I shall respond with 6; if you open with 6, I shall respond with 4. A few similar strategies are known.

> After {8,12} the pairs
> $(2, 3), (5, 7), (9, 11), \ldots, (4k + 1, 4k + 3)$
> and
> $(4, 6), (10, 14), (18, 22), \ldots, (8k + 2, 8k + 6)$,
> for $k \geq 1$, are mates.

There is a slightly more complicated strategy showing that another good reply to 6 is 9.

> After {6,9} mate the pairs
> (4,11), (5,8), (7,10) and $(3k + 1, 3k + 2)$ for
> $k \geq 4$,
> but *then*
> after 4,11 mate 5 with 7
> after 5,8 mate 4 with 7
> after 7,10 mate 4 with 5, 8 with 11.

We have proved that

$$\{2,3\} \ \{4,6\} \ \{6,9\} \ \{8,12\}$$
are all $\mathcal{P}$-positions,

and so

$$\{1\} \ \{2\} \ \{3\} \ \{4\} \ \{6\} \ \{8\} \ \{9\} \ \{12\}$$
are all $\mathcal{N}$-positions.

The numbers 1,2,3,4,6,8,9 and 12 are the only first moves for which explicit strategies have been found. You might expect that pairs (2,3), $(4k+1, 4k+3)$, (4,6), $(8k+2, 8k+6)$, (8, 12), $(16k+4, 16k+12)$ provide a strategy after $\{16,24\}$ but unfortunately 12 is *not* a legal move from the position $\{16, 24, 5, 7, 8\}$. On the other hand, for the strategies given above, both members of a pair are legal whenever one is. In fact 8 is a good reply to $\{16,24,5,7\}$ because it makes 16 and 24 irrelevant and we shall soon see that

$$\{5,7,8\} \ is \ a \ \mathcal{P}\text{-}position.$$

We don't know whether 24 is a good reply to 16, nor even whether 16 *has* any good reply.

Are All Openings Bad?

If on observing the fate of 1, 2 and 3 you thought maybe that all openings were bad, then probably our discussions of 4, 6, 8, 9 and 12 have tended to confirm your suspicions. In this section we'll try to analyze 5 and 7. The discussion of possible replies is made a lot easier by the **clique technique**.

You've already seen some cliques: The number 1 forms a rather special clique all by itself; 2 and 3 form another because they exclude all larger numbers. In our discussion of $\{4,5\}$, 6 and 7 formed a clique since they excluded 11. Cliques have the property that any reply to a clique member must also be a clique member and these two numbers must together exclude all numbers outside the clique.

We illustrate the clique technique by discussing $\{6,7\}$ (Fig. 3).
As usual, we can disregard 1, 2 and 3 which form the innermost cliques in *every* position. Now in Fig. 3, 4 and 5 *together* exclude all larger numbers and so form a third clique. *No matter what larger numbers have been named*, 4 will answer 5 and 5 will answer 4. We can therefore afford to neglect them in discussing larger numbers.

Now we assert that 8, 9, 10 and 11 form the next clique, because 8 and 10 together exclude all but 9 and 11, and these together exclude all but 8 and 10. Even when some larger numbers have already been named, 8 will answer 10, 9 will answer 11, and vice versa, and we can dismiss all four from the subsequent discussion.

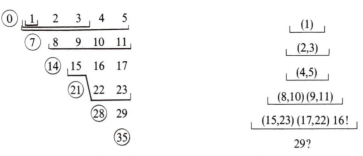

Figure 3. The Cliques after (6, 7).

We now know that any good reply to any of the remaining numbers

$$
\begin{array}{ccc}
15 & 16 & 17 \\
 & 22 & 23 \\
 & & 29
\end{array}
$$

must be another of these. We see that 15 answers 23 and vice versa since these leave only 16 and 17. Similarly 17 and 22 are mates. But since 16 excludes *both* 22 and 23, leaving only 15 and 17, it's a good move by itself. These five numbers form a clique, since 29 is always excluded.

> 16 is the unique
> good reply to {6,7}.

Table 6 in the Extras exhibits complete strategies in a similar way for all the positions

$$
\begin{array}{cccc}
\{4,5\}, & \{4,7\}, & \{4,9\}, & \\
\{5,6\}, & \{5,7\}, & \{5,8\}, & \{5,9\}, \\
\{6,7\}, & \{7,8\}, & \{7,9\}. &
\end{array}
$$

In particular it shows that

> $$
> \begin{array}{ccc}
> \{4,5,11\}, & \{4,7,13\}, & \{4,9,19\}, \\
> \{5,6,19\}, & \{5,7,8\}, & \{5,9,31\}, \\
> \{6,7,16\}, & \{7,9,19\}, & \{7,9,24\},
> \end{array}
> $$
> are $\mathcal{P}$-positions.

We deduce that any good reply to 5 or 7 must be at least a two-digit number. The smallest two digit number, 10, isn't a legal answer to 5; is it a good answer to 7? No!

> {7,10,12} is a $\mathcal{P}$-position.

This is proved by Fig. 4. Since the clique technique isn't as helpful as it might have been, we've added extra notes for three of the pairs.

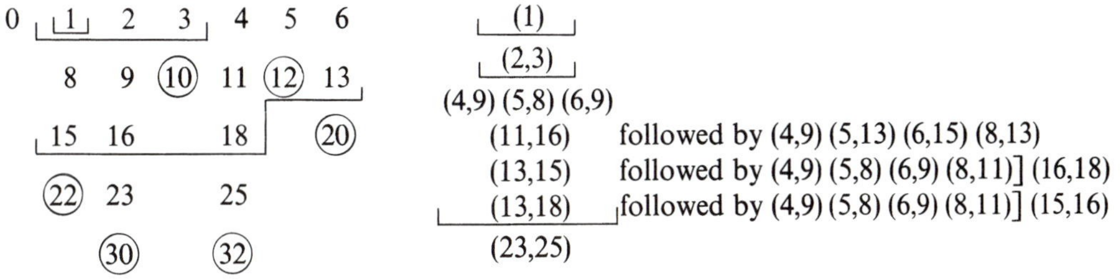

0	⌊1⌋	2	3	4	5	6
	8	9	⑩	11	⑫	13
	⌊15	16		18	⌋	⑳
	㉒	23		25		
		㉚		㉜		

⌊ (1) ⌋	
⌊ (2,3) ⌋	
(4,9) (5,8) (6,9)	
(11,16)	followed by (4,9) (5,13) (6,15) (8,13)
(13,15)	followed by (4,9) (5,8) (6,9) (8,11)] (16,18)
(13,18)	followed by (4,9) (5,8) (6,9) (8,11)] (15,16)
(23,25)	

Figure 4. The Position $\{7, 10, 12\}$.

Not All Openings Are Bad

R. L. Hutchings has proved that there can't be any good replies to 5 or 7! His main theorem is

> If a and b are coprime ($g = 1$)
> and $\{a, b\} \neq \{2, 3\}$, then $\{a, b\}$
> is an $\mathcal{N}$-position.

From this he deduces his **p-theorem**:

> If $p \geq 5$ is a prime number,
> $\{p\}$ is a $\mathcal{P}$-position,

p-positions are $\mathcal{P}$-positions

(for any legal reply produces a position with a g.c.d. of 1.) And from the *p*-theorem he deduces in turn his **n-theorem**:

> If n is a composite number
> not of the form $2^a 3^b$, then
> $\{n\}$ is an $\mathcal{N}$-position,

n-positions are $\mathcal{N}$-positions

(since n has a prime divisor $p \geq 5$, which is a good reply.) Together these account for the first few missing numbers:

$\{5\}, \{7\}, \{11\}, \{13\}, \{17\}, \ldots$ are $\mathcal{P}$-positions.
$\{10\}, \{14\}, \{15\}, \{20\}, \{21\}, \ldots$ are $\mathcal{N}$-positions.

Our explicit strategies accounted for the eight smallest numbers $2^a 3^b$:

$\{1\}, \{2\}, \{3\}, \{4\}, \{6\}, \{8\}, \{9\}, \{12\}$
are $\mathcal{N}$-positions.

But

Nobody knows about
$\{16\}, \{18\}, \{24\}, \{27\}, \{32\}, \{36\}, \ldots$!

(We'd be glad to be proved wrong.)

Strategy Stealing

Hutchings proves his main theorem by a fine piece of strategy stealing. He considers the topmost number, t, that is not excluded by $\{a, b\}$ and proves that if t is *not* a good reply, then some other number *is*!

We shall call $\{a, b\}$ an **end-position** because, as we'll see in a moment, the topmost number is excluded by every other legal move.

Now let's ask:

Is t a good reply to $\{a, b\}$?

If the answer is "yes", then $\{a, b\}$ is an $\mathcal{N}$-position.

If the answer is "no", then either the game is over or there is a good reply s to $\{a, b, t\}$. But since a, b and s exclude t, s is itself a good reply to $\{a, b\}$. We can say that the player to move from $\{a, b\}$ finds his strategy by stealing the second player's strategy, if he has one, for $\{a, b, t\}$.

In some cases, e.g. $\{5,9\}$, t (here 31) is a good reply. But in others, e.g. $\{5,7\}$ (where $t = 23$) it *isn't*. The strategy stealing argument only tells us that good moves exist, not what they are. Theft is no substitute for honest toil!

In general,

An end position with $t > 1$
is an $\mathcal{N}$-position,

end-positions are $\mathcal{N}$-positions.

But the end-position $\{2,3\}$ is *not* an $\mathcal{N}$-position. This is because $t = 1$ and the only legal move ends the whole game.

Why is $\{a,b\}$ an end-position if its g.c.d. is 1? In Fig. 5 we illustrate with $\{9,11\}$ for which the authors once knew no good reply (but see Table 5).

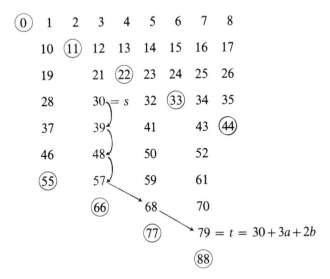

Figure 5. Hutchings's Theorem for $a = 9$, $b = 11$.

Writing the numbers in a columns, as is our wont, we see that in each column the *first excluded* (circled) number is a multiple of b, so the *last included* numbers must differ by multiples of b. Now from any legal move s we can get to the last legal number in its column by adding a's and from this we can get to t by adding b's showing that s excludes t (e.g. $s = 30$ in Fig. 5). The argument also provides a proof of Sylvester's well-known formula.

$$t = (a-1)b - a = ab - (a+b).$$

Quiet Ends

Suppose Hi and Lo have named two coprime numbers a and b and Hi is considering making the move s. Then we know that the topmost number t will be obtainable using sufficiently many coins of values s, a and b. But our argument proved that only *one* copy of the new coin will be needed:

$$t = s + ma + nb.$$

More generally from a position $\{a, b, c, \ldots\}$ we shall say that s **quietly** excludes t if t can be made up using any numbers of $a, b, c, \ldots$ together with *just one* copy of s:

$$t = ma + nb + \ldots + s.$$

A **quiet end-position** is one in which the topmost legal move is *quietly* excluded by every number not already excluded.

> If a is coprime with each of
> b and b_1, then
> $$S = \{a, bc, bd, be, \ldots\}$$
> is a quiet end-position
> if and only if
> $$S_1 = \{a, b_1c, b_1d, b_1e, \ldots\} \text{ is.}$$

THE QUIET END THEOREM

Thus

$$\{7, 1 \times 3, 1 \times 4\},$$

which is really the same position as $\{3,4\}$, is a quiet end-position , so that

$$\{7, 9, 12\} = \{7, 3 \times 3, 3 \times 4\}$$

and

$$\{7, 15, 20\} = \{7, 5 \times 3, 5 \times 4\}$$

are. In particular, these are end-positions and so are $\mathcal{N}$-positions by the strategy stealing argument. As usual we aren't told what the good replies are.

We shall use $\{7,9,12\}$ and $\{7,15,20\}$ to illustrate our proof of the quiet end theorem. Once again we write out the numbers in a (here 7) columns and circle the first excluded number in every column (Fig. 6). We assert that these numbers for the positions S and S_1 are in the proportion $b{:}b_1$ (3:5 in the example; see Fig. 7).

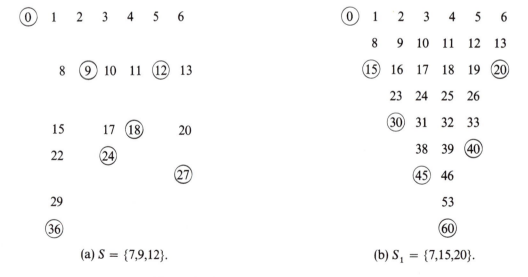

(a) $S = \{7,9,12\}$.   (b) $S_1 = \{7,15,20\}$.

Figure 6. Circled Numbers in Proportion.

$$\begin{array}{ccccccc}
& 29 & 2 & 17 & 11 & 5 & 20 \\
S: & \textcircled{0} & \textcircled{36} & \textcircled{9} & \textcircled{24} & \textcircled{18} & \textcircled{12} & \textcircled{27}
\end{array}$$

in the proportion

3

to

$$\begin{array}{ccccccc}
& 53 & 8 & 33 & 23 & 13 & 38 \\
S_1: & \textcircled{0} & \textcircled{60} & \textcircled{15} & \textcircled{40} & \textcircled{30} & \textcircled{20} & \textcircled{45}
\end{array}$$

5

Figure 7. The Circled Numbers Sorted.

We first see that the circled numbers for S really are multiples of b. Recall that we **circle** n for S if n is excluded by S, but $n - a$ is not. Since n is excluded, it has the form

$$n = ak + bm$$

where m would be excluded by $\{c, d, e, \ldots\}$. But if k were positive,

$$n - a = a(k - 1) + bm$$

would also be excluded by S, so $k = 0$ and we have simply

$$n = bm.$$

Now our assertion is that bm is circled for S only if $b_1 m$ is circled for S_1. Now $b_1 m$ is certainly *excluded* by S_1 and so is circled unless

$$b_1 m - a$$

is also excluded. But then we must have

$$b_1 m - a = ak + b_1 m'$$

for some m' excluded by $\{c, d, e, \ldots\}$, and

$$b_1 m = a(k + 1) + b_1 m',$$

showing that b_1 divides $k + 1$ since it is coprime with a. We can now divide by b_1 and multiply by b to obtain

$$bm = ak' + bm'$$

for some positive number k', showing that

$$bm - a = a(k' - 1) + bm'$$

was excluded, and bm was *not* circled for S.

In its modest way, the quiet end theorem is quite powerful. It often gives the quietus to infinitely many replies with a single blow.

No odd number is a
good reply to $\{16, 24\}$.

For 1 clearly isn't and if a is any other odd number then $\{a, 2, 3\}$ is really the same as the quiet end position $\{2,3\}$. By the quiet end theorem $\{a, 16, 24\}$ is a quiet end-position and so an $\mathcal{N}$-position.

In a similar way it proves that $\{4,6\}$ and $\{6,9\}$ are $\mathcal{P}$-positions without bothering to provide a detailed strategy. Let's use it to discuss the position $\{8,10\}$. After $\{4,5\}$ we found that the only remaining moves were

$$1, \quad 2, \quad 3, \quad 6, \quad 7, \quad 11,$$

so after $\{8,10\}$ the only remaining even numbers will be twice these,

$$2, \quad 4, \quad 6, \quad 12, \quad 14 \quad 22.$$

The quiet end theorem enables us to say that any good reply to $\{8,10\}$ must be in one of these two sets, for otherwise it is an odd number a excluded by $\{4,5\}$ so $\{a, 4, 5\}$ and therefore $\{a, 8, 10\}$ will be quiet end-positions. Now,

1	loses instantly,
(2,3)	are mated as usual,
(4,6)	eliminate 8,10 and will mate, as will
(7,11)	(see $\{6,7\}$ in Table 6 in the Extras) and
(12,14)	by our strategy for $\{8,12\}$.

So 22 is the only hope for a good reply to $\{8,10\}$. We shall see later that

$$\boxed{\{8,10,22\} \text{ is a } \mathcal{P}\text{-position.}}$$

Doubling and Tripling?

Note that the $\mathcal{P}$-position $\{8,10,22\}$ is the *double* of $\{4,5,11\}$. Our $\{8,12\}$ strategy shows that all $\mathcal{P}$-positions arising in the $\{4,6\}$ strategy have doubles that are also $\mathcal{P}$-positions. Maybe every $\mathcal{P}$-position doubles to another? No! For $\{5,6,19\}$ is $\mathcal{P}$, but $\{10,12,38\}$ is answered by 7 since $\{10,12,38,7\}$ is really the same as $\{7,10,12\}$.

Maybe the *triple* of every $\mathcal{P}$-position is another? No! This time $\{4,5,11\}$ is $\mathcal{P}$, but $\{12,15,33\}$ is answered by 5 since $\{5,12,33\}$ is a $\mathcal{P}$-position, as we'll soon see.

Halving and Thirding?

Nevertheless there are many $\mathcal{P}$-positions whose doubles and triples are still $\mathcal{P}$. We conjecture:

¿If $\{2a, 2b, 2c, \ldots\}$ is $\mathcal{P}$ so is $\{a, b, c, \ldots\}$?	and	¿If $\{3a, 3b, 3c, \ldots\}$ is $\mathcal{P}$ so is $\{a, b, c, \ldots\}$?

Finding the Right Combinations

How should you start a game of Sylver Coinage? Now that you know so much you will perhaps name 5 for your first move. You now have a strategy for every move I might make and probably feel a little safe. But those stolen strategies are firmly locked inside that little safe you're feeling and more than sensitive fingers are needed to find the right combinations.

You know the first few: 1 needs no reply and you should make the pairs (2,3), (4,11), (6,19), (7,8) and (9,31). Is there any general rule? In trying to answer this question for you we went to a lot of trouble and eventually found a fairly efficient way of breaking open the safe. But the winning combinations it reveals (Fig. 8) suggest that there is no simple answer.

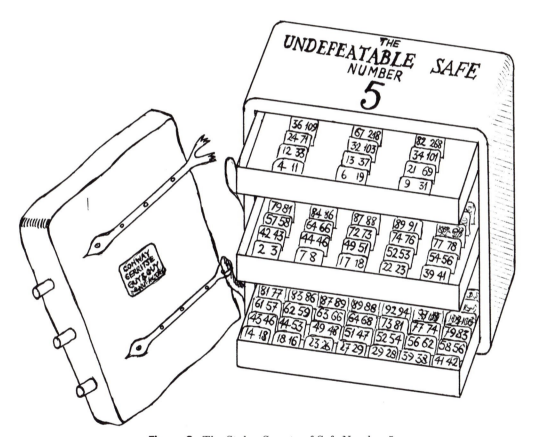

Figure 8. The Stolen Secrets of Safe Number 5.

Let's take a closer look at a position in which 5 and some other numbers have been named. If we were to write the numbers in five columns as usual we would circle 0 and just four other numbers a, b, c, d in the 1-, 2-, 3-, 4-columns respectively, as in Fig. 9. We now make a three-dimensional table of $\mathcal{P}$-positions using just three of these numbers as headings and the fourth as an entry.

Table 1(a) shows the case in which a is the entry and b, c, d the row, column and layer headings. Tables 1(b,c,d) have b, c, d as entries.

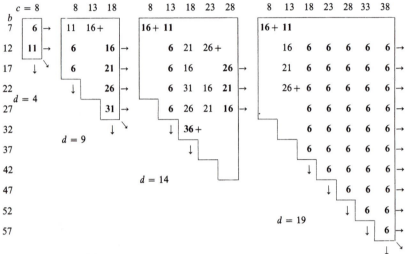

Table 1(a). Entries a for $\mathcal{P}$-positions $\{5, a, b, c, d\}$.

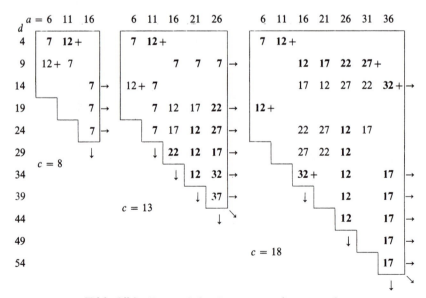

Table 1(b). Entries b for $\mathcal{P}$-positions $\{5, a, b, c, d\}$.

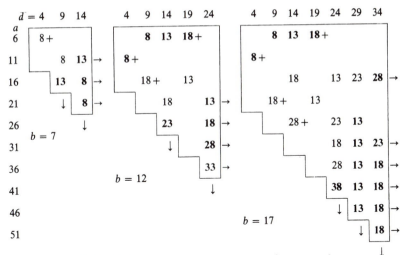

Table 1(c). Entries c for $\mathcal{P}$-positions $\{5, a, b, c, d\}$.

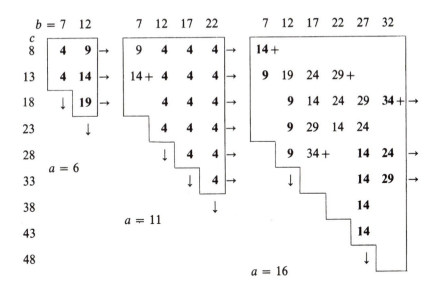

Table 1(d). Entries d for $\mathcal{P}$-positions $\{5, a, b, c, d\}$.

Some positions will appear repeatedly because a heading is redundant. These are indicated by bold figures. For example

$$\{5, 6, 12, 13, 14\}, \{5, 6, 17, 13, 14\}, \{5, 6, 22, 13, 14\}, \ldots$$

are really the same position because $12 = 6 + 6$ is redundant and so we have a column of **6**'s in layer 14 of Table 1(a). In {5,6,12,18,19} both 12 and 18 are redundant, so the 19 layer of that table is almost entirely made up of **6**'s.

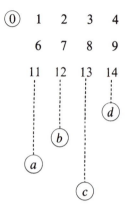

Figure 9. The General Position $\{5, x, y, \ldots\}$.

In $\{5, 16, 7, 13, 9\}$ it is the *entry* $16 = 7 + 9$ that is redundant, so 16 can be replaced by any of

$$21, 26, 31, 36, 41, \ldots$$

and we have written 16+ to indicate this. Really an entry $n+$ is short for infinitely many entries

$$n, n + 5, n + 10, n + 15, n + 20, \ldots$$

The entries in Table 1(a) were computed in lexicographic order, by making due allowance for these repetitions and otherwise entering the least number $5k + 1$ not appearing earlier in the same row, column or file.

You'll probably find the method easier to follow in Table 2 which deals with positions $\{4, a, b, c\}$ in a similar way. This time each entry is the smallest number $b = 4k + 2$ which has not appeared earlier in its row or column and an entry $b+$, shorthand for

$$b, b + 4, b + 8, b + 12, b + 16, \ldots$$

is made when $b = 2a$ or $2c$. It can be deduced from the quiet end theorem that other kinds of repetition will not appear.

Table 3 gives pairs $x < y$ for which $\{4, x, y\}$ is already a $\mathcal{P}$-position, extracted from an extended version of Table 2, kindly calculated for us by Richard Gerritse. It seems that the ratio y/x approaches $2 \cdot 56 \ldots$

As soon as 4 or 5 arises in your game you should refer to the appropriate one of these tables. If 6 turns up first, see the corresponding Table 7 in the Extras.

a	7	11	15	19	23	27	31	35	39	43	47	51	55	59	63	67	71	75	79	83	87	91
5	6	10+																				
9	10	6	14	18+																		
13	14+		6	10																		
17		10	6	14	18	22	26	30	34+													
21		18	14	6	10	26	22	34	30	38	42+											
25		22		10	6	14	18	26			30	34	38	42	46	50+						
29		26		18	14	6	10	22			34	30	42	38	50	46	54	58+				
33		30+		22	26	10	6	14	18													
37				26	22	18	14	6	10	42	38	30	34	54		46	50	58	62	66	70	
41				30	34	38	42	10	6	14	18	22	26	58		50	46	54	66	62	74	
45				34	30	42	38	18	14	6	10	26	22	62		58	54	46	50	70	66	
49				38	42	30	34	46	22	10	6	14	18	26		62		50	54	58	78	
53				42	38	34	30	50	26	18	14	6	10	22		66		62	46	54	58	
57						46+			38		22	26	10	6	14	18	30	34	42			
61						46	50	54	42		26	22	18	14	6	10	34	30	38	58	74	62
65						50	46	58	54		62		34	30	10	6	14	18	22	26	38	42
69							54+		46		50			18	14	6	10	26	22	30	34	
73							54	50	58		46		62	66	30	22	10	6	14	18	26	38
77							58	62	66		54		46	50	34	26	18	14	6	10	22	30
81							62+				58		50	46	38	30	22	26	10	6	14	18
85								66	62		70		54	58	42	34	26	22	18	14	6	10
89								70+			66		58	54		38	42		30	34	10	6
93									70		74		66	62	78	42	38		34	30	18	14
97									74		78		70	82	66		86	38	90	42	34	22
101									78+				74	70				42	66	38	46	26
105											82		78	74	70		90		86	94	42	46
109											86		82	78	74		70		94	90	50	54
113											90		86	94	82		74		70	78	98	50
117											94+		90	86			78		74	70	82	

Table 2. Values of b for which $\{4, a, b, c\}$ is a $\mathcal{P}$-position.

x	y	x	y	x	y	x	y	x	y	x	y	x	y
5	11	107	269	205	531	303	777	405	1043	501	1291	603	1549
7	13	109	279	207	529	305	783	407	1045	503	1289	605	1555
9	19	111	277	211	541	311	797	409	1051	505	1299	607	1557
15	33	113	287	213	547	313	807	411	1053	511	1309	609	1567
17	43	119	301	219	557	315	805	413	1063	513	1319	615	1577
21	51	121	307	221	567	317	819	415	1065	519	1329	617	1583
23	57	123	309	223	569	321	823	419	1077	521	1339	621	1595
25	67	125	319	227	585	323	829	421	1079	523	1341	623	1593
27	69	129	331	229	583	325	839	425	1095	525	1347	625	1603
29	75	131	333	231	593	327	841	427	1097	527	1349	627	1609
31	81	133	343	233	595	329	851	429	1103	533	1371	633	1627
35	89	135	345	237	611	335	857	431	1105	535	1373	635	1625
37	95	141	363	239	613	337	871	433	1115	537	1379	637	1635
39	101	143	365	241	619	339	869	439	1125	539	1381	639	1637
41	103	147	373	245	631	341	879	441	1135	543	1393	643	1649
45	115	149	379	247	629	347	885	447	1145	545	1395	645	1655
47	117	151	385	251	641	349	899	449	1151	549	1411	651	1665
49	127	153	391	253	647	351	901	451	1157	551	1413	653	1679
53	139	155	397	255	649	353	911	455	1165	553	1423	655	1681
55	137	157	399	257	659	355	909	457	1171	555	1425	657	1687
59	145	163	417	259	665	357	919	459	1177	559	1433	661	1699
61	159	165	423	261	671	359	917	461	1183	561	1443	663	1697
63	161	169	435	265	683	361	927	463	1189	563	1445	667	1709
65	167	171	437	267	685	367	941	465	1195	565	1451	669	1719
71	177	173	443	271	697	369	951	469	1207	571	1465	673	1731
73	183	175	445	273	699	371	953	471	1209	573	1471	675	1729
77	195	179	453	275	705	375	961	473	1215	575	1477	677	1739
79	193	181	467	281	723	377	971	479	1225	577	1483	679	1741
83	209	185	475	283	725	381	983	481	1235	579	1485	681	1751
85	215	187	477	285	731	383	981	485	1247	581	1495	687	1757
87	217	189	483	289	743	387	993	487	1245	587	1505	689	1771
91	225	191	489	291	745	389	1003	491	1257	589	1511	691	1769
93	235	197	507	293	755	393	1011	493	1267	591	1517	693	1783
97	243	199	509	295	757	395	1013	495	1269	597	1531	695	1781
99	249	201	515	297	767	401	1031	497	1279	599	1537	701	1803
105	263	203	517	299	765	403	1029	499	1277	601	1543	703	1801
												707	1813

Table 3. Pairs x, y for which $\{4, x, y\}$ is a $\mathcal{P}$-position.

What Shall I Do When g Is Two?

Example: $\{8, 10, 22\}$

Apparently we have to examine infinitely many possible replies. Fortunately there is a way of doing this in a finite time. A similar method will work for *any* position with $g = 2$.

Let's see how the position will look after some play from $\{8, 10, 22\}$. If 1, 2 or 3 has been played, we know what to do. Otherwise the only even numbers that can have been played are

4, 6, 12, 14

and the even part of the position must look like one of

$$\{4,6\} \; \{4,10\} \; \{6,8,10\}$$
$$\{8,10,12,14\} \; \{8,10,12\} \; \{8,10,14\} \; \{8,10,22\}$$

What odd numbers have been played? If the least of them is n, then since $\{8,10,22\}$ excludes all of

$$16, 18, 20, 22, 24, \ldots$$

we can suppose that the only relevant odd numbers are among

$$n, n+2, n+4, n+6, n+12, n+14.$$

And if any even moves have been made they will restrict the possibilities still further. For instance if 6 has been played we can suppose that the odd numbers are one of the sets

$$n, \, n+2, \, n+4 \qquad n, \, n+2 \qquad n, \, n+4 \qquad n$$

Table 4 shows the status of the positions classified in this way. Since the last four columns repeat indefinitely, this finite table contains the information for every odd n. How was it computed and why is it periodic?

Let's take a typical entry:

$$\{8, 10, 14, n, n+6, n+12\}.$$

From this position there are three kinds of option:

(a) the *even* numbers 4, 6 or 12,
(b) the *small odd* numbers $m \le n - 14$,
(c) the *large odd* numbers $n - 12, n - 10, n - 8, n - 6, n - 4, n - 2, n + 2, n + 4$.

Case (a) leads to a position (in an earlier segment of the table) with even part

$$\{4, 10\}, \{6, 8, 10\} \text{ or } \{8, 10, 12, 14\}$$

and we can suppose that these have already been analyzed and found to be ultimately periodic in n.

A case (b) move leads to

$$\{8, 10, 14, m\}$$

since m excludes n and all larger odd numbers. If there is any odd m for which this is a $\mathcal{P}$-position, then $\{8, 10, 14, n, n+6, n+12\}$ will be an $\mathcal{N}$-position for all $n \ge m + 14$. If not, we can reject moves in case (b).

Finally, case (c) moves either leave n unchanged or decrease it by at most 12. We conclude that the outcome of every position in the table is computed in a fixed way from

> ultimately periodic information (case (a)),
> ultimately constant information (case (b)), and
> information in the last few columns (case (c));

it must therefore be ultimately periodic in n.

Set	Combo	5	7	9	11	13	15	17	19	21	23	25	27	29	31	33	35	37	39	41	43	45	47	49	51
{4,6} (P)	$n, n+2$	P	–	P	–	P	–	P	–	P	–	P	–	P	–	P	–	P	–	P	–	P	–	P	–
	n	–	–	–	–	–	–	–	–	–	–	–	–	–	–	–	–	–	–	–	–	–	–	–	–
{4,10} (N)	$n, n+2$	–	P	–	–	P	–	–	P	–	–	P	–	–	P	–	–	P	–	–	P	–	–	P	–
	$n, n+6$	P	–	–	–	P	–	–	–	P	–	–	–	P	–	–	–	P	–	–	–	P	–	–	–
	n	–	–	–	–	–	–	–	–	–	–	–	–	–	–	–	–	–	–	–	–	–	–	–	–
{6,8,10} (N)	$n, n+2, n+4$	–	P	–	P	–	–	–	–	–	–	–	–	–	–	–	–	–	–	–	–	–	–	–	–
	$n, n+2$	–	–	–	–	–	–	–	–	–	–	–	–	–	–	–	–	–	–	–	–	–	–	–	–
	$n, n+4$	P	–	–	–	–	–	–	–	–	–	–	–	–	–	–	–	–	–	–	–	–	–	–	–
	n	–	P	–	P	–	–	–	–	–	–	–	–	–	–	–	–	–	–	–	–	–	–	–	–
{8,10,12,14} (P)	$n, n+2, n+4, n+6$	P	–	P	–	P	–	P	–	P	–	P	–	P	–	P	–	P	–	P	–	P	–	P	–
	$n, n+2, n+4$	–	–	–	–	–	–	–	–	–	–	–	–	–	–	–	–	–	–	–	–	–	–	–	–
	$n, n+2, n+6$	–	–	–	–	–	–	P	–	–	–	P	–	–	–	P	–	–	–	P	–	–	–	P	–
	$n, n+2$	P	–	P	–	P	–	P	–	P	–	P	–	P	–	P	–	P	–	P	–	P	–	P	–
	$n, n+4, n+6$	–	P	–	P	–	–	–	P	–	–	–	P	–	–	–	P	–	–	–	P	–	–	–	P
	$n, n+4$	–	–	P	–	–	P	–	–	–	P	–	–	–	P	–	–	–	P	–	–	–	P	–	–
	$n, n+6$	–	–	P	–	–	P	–	–	–	P	–	–	–	P	–	–	–	P	–	–	–	P	–	–
	n	–	–	–	–	–	–	–	–	–	–	–	–	–	–	–	–	–	–	–	–	–	–	–	–
{8,10,12} (N)	$n, n+2, n+4, n+6$	P	–	–	–	–	–	–	–	–	–	–	–	–	–	P	–	–	–	P	–	–	–	–	P
	$n, n+2, n+4$	–	–	P	–	–	–	P	–	–	–	–	–	–	–	–	–	–	–	–	–	P	–	–	–
	$n, n+2, n+6$	–	–	P	–	–	–	–	–	–	–	–	–	–	–	–	–	P	–	–	–	P	–	–	–
	$n, n+2$	P	–	–	–	–	–	–	–	–	–	–	–	–	–	–	–	P	–	–	–	–	–	–	P
	$n, n+4, n+6$	–	P	P	–	–	P	–	–	–	–	–	–	–	–	–	–	–	–	–	–	P	–	–	P
	$n, n+4$	–	–	–	P	–	–	–	P	–	–	P	–	–	P	P	–	–	P	P	–	–	P	–	P
	$n, n+6$	–	–	–	–	–	–	P	–	–	–	P	–	–	P	–	–	–	P	–	–	–	P	–	–
	$n, n+14$	–	–	P	–	–	P	–	–	–	–	–	–	–	–	–	–	–	–	–	–	–	–	–	–
	n	–	–	–	–	–	–	–	–	–	–	–	–	–	–	–	–	–	–	–	–	–	–	–	–
{8,10,14} (N)	$n, n+2, n+4, n+6$	P	–	–	–	–	–	–	–	–	–	–	–	–	–	–	P	–	–	–	–	–	–	–	–
	$n, n+2, n+4$	–	–	P	–	–	–	–	–	–	–	–	–	P	–	P	–	–	–	–	–	–	–	–	–
	$n, n+2, n+6$	–	–	P	–	–	–	–	–	–	–	–	–	P	–	P	–	–	–	–	–	–	–	–	–
	$n, n+2$	–	–	–	–	–	–	–	–	–	–	–	–	–	–	–	–	–	–	–	–	–	–	–	–
	$n, n+4, n+6$	–	–	P	–	–	P	–	–	–	–	–	–	P	–	–	–	P	–	–	–	–	–	–	–
	$n, n+4$	–	P	–	–	–	–	–	P	–	–	–	–	–	–	–	–	–	–	–	–	–	P	–	–
	$n, n+6, n+12$	–	–	–	–	–	–	P	–	–	–	–	–	–	–	–	–	P	–	P	–	–	–	–	–
	$n, n+6$	–	–	–	–	–	–	–	–	–	–	–	–	–	–	–	–	–	–	–	–	–	–	–	–
	$n, n+12$	P	–	P	–	P	P	–	–	–	–	–	–	P	–	P	–	P	–	–	–	–	–	–	–
	n	–	–	–	–	–	–	–	–	–	–	–	–	–	–	P	–	P	–	–	–	–	–	–	–
{8,10,22} (P)	$n, n+2, n+4, n+6$	P	–	P	–	P	–	P	–	P	–	P	–	P	–	P	–	P	–	P	–	P	–	P	–
	$n, n+2, n+4$	–	–	–	–	–	–	–	–	–	–	–	–	–	–	–	–	–	–	–	–	–	–	–	–
	$n, n+2, n+6$	–	–	–	P	–	–	–	–	–	–	–	P	–	–	–	P	–	–	–	P	–	–	–	P
	$n, n+2, n+14$	–	–	P	–	P	–	P	–	P	–	P	–	P	–	P	–	P	–	P	–	P	–	P	–
	$n, n+2$	P	–	–	–	–	–	–	–	–	–	–	–	–	–	–	–	–	–	–	–	–	–	–	–
	$n, n+4, n+6$	–	–	–	–	–	–	–	–	–	–	–	–	–	–	–	–	–	–	–	–	–	–	–	–
	$n, n+4$	–	P	P	–	–	P	P	–	–	P	P	–	–	P	P	–	–	P	P	–	–	P	P	–
	$n, n+6, n+12$	–	–	P	–	–	–	P	–	–	–	P	–	–	–	P	–	–	–	P	–	–	–	P	–
	$n, n+6$	–	–	–	–	–	–	–	–	–	–	–	–	–	–	–	–	–	–	–	–	–	–	–	–
	$n, n+12, n+14$	–	–	–	–	–	–	–	–	–	–	–	–	–	–	–	–	–	–	–	–	–	–	–	–
	$n, n+12$	–	–	–	–	–	–	–	–	–	–	–	–	–	–	–	–	–	–	–	–	–	–	–	–
	$n, n+14$	P	–	–	–	P	–	–	–	–	–	–	–	–	–	–	–	–	–	–	–	–	–	–	–
	n	–	–	–	–	–	–	–	–	–	–	–	–	–	–	–	–	–	–	–	–	–	–	–	–

Table 4. The Position $\{8, 10, 22\}$.

Every position with $g = 2$ can be handled in this way. When we have computed enough to verify the period, we can decide in particular whether there is any good reply. For $\{8,10,22\}$ there isn't one, so it is a $\mathcal{P}$-position.

The Great Unknown

We can best describe our knowledge in terms of the number g. When

$$g = 1$$

the position is bounded so you can find what to do by working through all positions. Of course this might take a long time even if one of our theorems already tells you the outcome. We know that there must a be good reply to $\{31,37\}$ but don't know any method which guarantees to find one in the next millenium. When

$$g = 2$$

the method we have just described will compute the outcome in a finite but probably even longer time. If

$$g \text{ is divisible by a prime } p \geq 5$$

then p is a good reply when it hasn't already been named; if it has, of course there isn't any.

The authors have only been able to examine a few particular positions with other values of g. Table 8 in the Extras contains a complete discussion of $\{6,9\}$. Although this is a two-dimensional table, a periodicity develops which enables us to analyze the position to infinity. Maybe a similar thing happens for some other positions with $g = 3$. We computed a much larger three-dimensional table for $\{8,12\}$ ($g = 4$), but could detect no structure outside the range covered by our explicit strategy.

16 is the first opening move whose status is in doubt. We don't know whether $\{16\}$ has a good reply nor even any way of finding out in any finite time. You might consider working upwards testing each possible reply in turn and hoping to detect some structure, but even this is impossible. We don't know any way to test the reply 24, say, in any finite time. We don't even know how to test 100, say, as a possible reply to $\{16,24\}$!

The quiet end theorem often eliminates infinitely many replies, for example all odd replies to $\{6\}$ or to $\{16,24\}$, but it never eliminates any reply that would be infinitely hard to analyze.

	{7,9,11}	{7,9}	{7,11}	{7}	{9,11}	{9}	{11}	{ }
{6,8,10}	[]	[11]	[9]	[]	[4, 5, 7]	[5]	[]	[4, 7, 11]
{6,8}	[10]	[]	[]	[9, 10, 11]	[4, 5]	[5, 7]	[7, 10]	[4]
{6,10}	[8]	[]	[15]	[8, 9]	[4]	[7]	[8, 13]	[4]
{6}	[]	[8, 10, 11]	[8, 9]	[16]	[4, 7]	[]	[26]	[4, 9]
{8,10,12}	[4, 5, 6]	[4]	[13]	[5, 6]	[13, 14, 15]	[23]	[6]	[14]
{8,12}	[5]	[6]	[6]	[5]	[]	[11, 15]	[9, 13]	[]
{10,12}	[4]	[4, 6]	[16]	[]	[]	[11, 13, 14]	[9, 14, 15, 17]	[7
{12}	[6]	[15]	[27]	[10]	[8, 10]	[6]	[49]	[8
{8,10}	[4, 5, 6]	[4]	[]	[5, 6, 11]	[23]	[13, 15]	[6, 7]	[22]
{8}	[5]	[6]	[6, 10]	[5]	[12]	[21]	[23]	[12
{10}	[4]	[4, 6]	[8]	[12]	[12]	[16]	[24, 28, 47]	[5
{ }	[6]	[19, 24]	[24, 34]	[]	[13, 30]	[6]	[]	[5, 7, 11, 13, …

Table 5. Status of Subsets of {6,7,8,9,10,11,12} and Known Good Replies.

Even members of set at Left; Odd members at head. Bracket is closed when *all* good replies are known, so that [] indicates a $\mathcal{P}$-position. The last entry contains all primes greater than 3, and *may* contain some entries $2^a 3^b$.

Table 5 tells you the outcome and all the good replies we know to every position made from the numbers

$$6, 7, 8, 9, 10, 11, 12$$

(if 4 or 5 is involved, Tables 2, 3, 1 and Fig. 8 go much further). If you can add any more to this table or decide whether any number $2^a 3^b$ is a good opening move we would like to hear from you.

Are Outcomes Computable?

We can prove that there *must be* a way of programming a computer to find the outcome of $\{n\}$ even though we don't know what that way *is*! The reason is:

> There can only be finitely
> many good opening moves
> $2^a 3^b$.

For no one of these can divide any other, so that no two can have the same value of a or the same value of b. So if $2^{a_0} 3^{b_0}$ is such a number with a_0 as small as possible, and $2^a 3^b$ is any other, then we must have $b < b_0$ and so there are at most $b_0 + 1$ such numbers, say $n_1, n_2, \ldots, n_k$. We suspect there are none!

If you only knew what these numbers were, then you could program your machine with PORN (Fig. 10) and work out the outcome of any $\{n\}$. This argument shows that in the purely technical sense this is a computable function of n, even though we don't know what function it is.

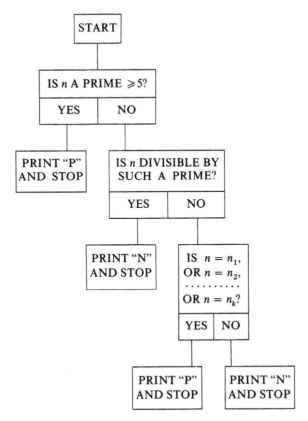

Figure 10. PORN, A Program which Decides if $\{n\}$ is $\mathcal{P}$ OR $\mathcal{N}$.

The Etiquette of Sylver Coinage

Few Western readers can understand the subtleties of etiquette in the oriental country from which our game comes. But at least we can save you from the more obvious gaffes by pointing out that in Sylver Coinage it is customary for a player who knows he is winning to resign by naming 1, 2 or 3. This quaint custom is said to originate in the tradition that Hi, who could see much further than Lo, nobly took upon himself the fate that was about to befall his beloved brother.

When it's plain to all the world that you have a win, any move but 1 will insult your opponent, but in other cases we advise you to name 3 (2 is possible, but may be misunderstood). If your opponent concurs in your analysis, he will respond with 2, but you have allowed him to express another opinion by naming 1. (Replies to 3 other than 1 or 2 may also be available but their nuances are harder to interpret.)

Of course, one of the greatest insults you can offer is to name 1, 2 or 3 at the very start of the game, for this is the philosopher Hu Tchings' prerogative, at least until someone finds a new way to win.

Extras

Chomp

Here is a game with similar rules to Sylver Coinage. For some fixed number N, the players alternately name divisors of N which may not be multiples of previously named numbers. Whoever names 1 loses. If $N = 432 = 2^4 3^3$, for example, a move is essentially to eat a square (e.g. 36) from the chocolate bar in Fig. 11, together with all squares below and/or to the right of it. Square number 1 is poisoned!

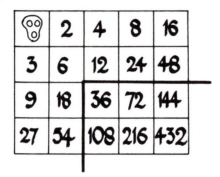

Figure 11. Chomping at a Chocolate Bar.

The first few $\mathcal{P}$-positions are shown in Fig. 12. Strategy stealing shows that rectangles larger than 1×1 are $\mathcal{N}$-positions; the replies are unique if either side is at most 3, but Ken Thompson found that 4×5 and 5×2 bites both answer 8×10.

The arithmetic form of the game is due to Fred. Schuh, the geometric one to David Gale.

Zig-Zag

Two players alternately name distinct numbers (which are allowed to be fractional or negative) and the game ends as soon as the resulting sequence contains either an increasing subsequence (**zig**) of length a or a decreasing one (**zag**) of length b. The normal play $a + 1$, $b + 1$ game is really the same as the misère a, b game, and so we consider only the latter.

Zig-Zag, which was suggested to us by S. Fajtlowicz, sounds difficult to analyze, but fortunately there is a rather clever transformation into a geometrical game like Chomp. We regard square (r, s) in Fig. 14 as eaten if the number sequence so far contains a rising zig of length r and a sagging zag of length s that end with the same number. Then the moves are as in Chomp except that the first move may eat square $(1, 1)$ only, and the innermost square eaten on any subsequent move must be adjacent to a previously eaten square. The squares $(a, 1)$ and $(1, b)$ are poisoned, so play really goes on inside the outlined $a - 1$ by $b - 1$ **chocolate bar** of Fig. 14.

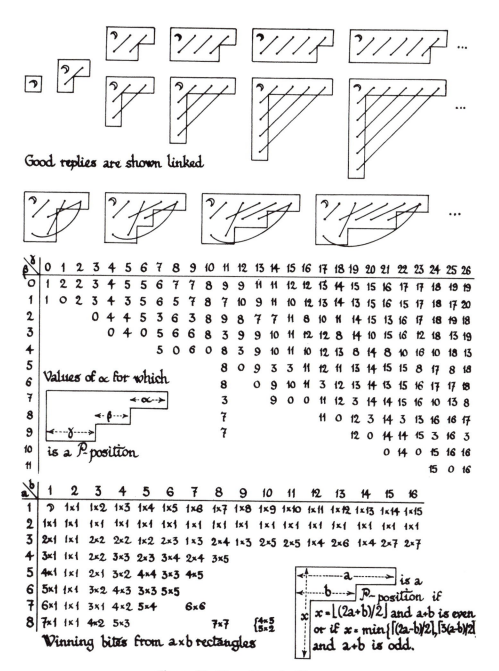

Figure 12. $\mathcal{P}$-positions in Chomp.

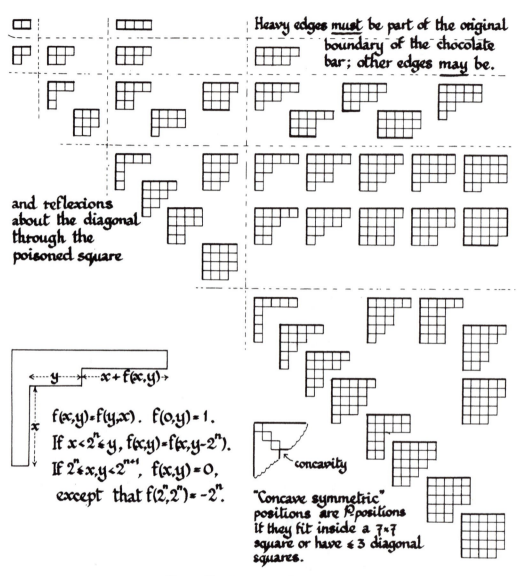

Heavy edges **must** be part of the original boundary of the chocolate bar; other edges **may** be.

and reflexions about the diagonal through the poisoned square

$f(x,y) = f(y,x)$. $f(0,y) = 1$.

If $x < 2^n \leqslant y$, $f(x,y) = f(x, y - 2^n)$.

If $2^n \leqslant x, y < 2^{n+1}$, $f(x,y) = 0$,

except that $f(2^n, 2^n) = -2^n$.

concavity

"Concave symmetric" positions are $\mathcal{P}$-positions if they fit inside a 7×7 square or have $\leqslant 3$ diagonal squares.

Figure 13. $\mathcal{P}$-positions for Zig-Zag.

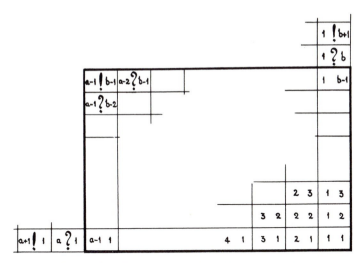

Figure 14. The Chocolate Bar Form of Zig-Zag.

If $a \geq 3$, $b \geq 3$ and $a + b \leq 17$, the first player wins the misère a-Zig, b-Zag game, because David Seal's calculations show that the corresponding $a - 1$ by $b - 1$ chocolate bars are $\mathcal{N}$-positions.

By assigning Heads to Horizontal edges and Tails to verTical ones we get an equivalent game with coin sequences, involving moves of a head rightwards over tails or a tail leftwards over heads, and Seal used this idea to compute Fig. 13 showing all $\mathcal{P}$-positions for which the uneaten part of the chocolate bar fits inside a 5×5 square.

To find $\mathcal{P}$-positions in both Chomp and Zig-Zag, we used the tabular technique of Chapter 15, and the Clique Technique of this one.

More Cliques for Sylver Coinage

To follow the cliques in Table 6, we advise you to set out the remaining numbers as we did in Figs. 2, 3, 4 for the cases {4,5}, {6,7} and {7,10,12}. Numbers not mentioned are excluded by a good reply.

5-Pairs

The safe combinations $\{5, x, y\}$ are of three types. In the top drawer in Fig. 8 are those with y so much larger than x that the coordinates $\{a, b, c, d\}$ are $\{x, 2x, 3x, y\}$. For these it seems that y/x tends to 3. But are there infinitely many numbers in the top drawer? The middle drawer contains the remaining ones for which $x + y$ is a multiple of 5. It seems that for these, x and y always differ by 1 or 2. In the bottom drawer we have arranged the pairs with coordinates $\{x, y, x + y, 2y\}$ where x and y are in the order given. It seems that here, as in the second drawer, y/x tends to 1.

position	replies	strategy, with cliques indicated by]
{4,5}	11!	1?](2,3)](6,7)]11!
{4,7}	13!	1?](2,3)](5,6)](9,10)]13!
{4,9}	19!	1?](2,3)](5,11)(6,11)(7,10)(14,15)]19!
{5,6}	19!	1?](2,3)](4,7)](8,9)](13,14)]19!
{5,7}	8!	1?](2,3)](4,6)(9,13)(11,13)8!
{5,8}	7!	1?](2,3)](4,11)(6,9)7!
{5,9}	31!	1?](2,3)](4,11)(6,8)(7,13)](12,16)](17,21)](22,26)]31!
{6,7}	16]	1?](2,3)(4,5)](8,9)(8,10)(8,11)(9,10)(9,11)](15,23)(17,22)16!
{7,8}	5!	1?](2,3)](4,13)(6,9)(6,10)(6,11)5!
{7,9}	19!24!	1?](2,3)](4,10)(5,13)(6,8)(6,10)(6,11)](12,15)(17,20)(22,26)(29,33)19!24!
		after {7,9,22,26} (12,17)(19,24)(15,20)
		{7,9,29,33} (12,15)(17,20)(19,31)(22,26)(24,26) ...
		{7,9,19} (12,15)(15,17)(15,20)](22,24)(29,31)
		{7,9,24} (12,15)(17,20)(19,22)](26,29)

Table 6. Some Complete Strategies for Sylver Coinage.

Positions Containing 6

As in our other analyses, we write the numbers in six columns and circle 0 and five other numbers a, b, c, d, e, one in each of the 1-, 2-, 3-, 4- and 5-columns respectively. We tabulate $\mathcal{P}$-positions by entries c in a 4-dimensional table (Table 7) whose coordinates, a, b, d, e are congruent to 1, 2, 4, 5, modulo 6.

Entries outside the areas enclosed by full lines are found by repeating entries according to the arrows, where appropriate. The tables for $b = 8$, $d = 4$ and $b = 8$, $d = 16$ can be extended indefinitely by repeating the portions between the pecked lines and increasing all entries by 12 or 60 respectively. The two tables with $d = 10$, $b = 8$ or 14 contain no further entries. All further entries in that for $b = 14$, $d = 16$ are **15**.

Table 8 represents a complete discussion of positions containing {6,9}. Reduced positions contain one, or possibly two neighboring, numbers of form $3k + 1$, and of form $3k + 2$; pairs of rows and columns refer to the latter and former possibilities. Positions represented by cells outside the crenellated line are not reduced. The pattern within the rectangular quadrant continues indefinitely. The minus signs denote $\mathcal{N}$-positions, the plus signs $\mathcal{P}$-positions and the "=" signs $\mathcal{N}$-positions which, at a casual glance at the pattern, might be mistaken for $\mathcal{P}$-positions.

Sylver Coinage Has Infinite Nim-Values

If we make naming 1 an *illegal* rather than a *stupid* move, Sylver Coinage becomes a normal play rather than a misère play game and we could consider adding it to other games using the Sprague-Grundy theory. However since some positions have infinitely many options, we can expect infinite nim-values and indeed they happen!

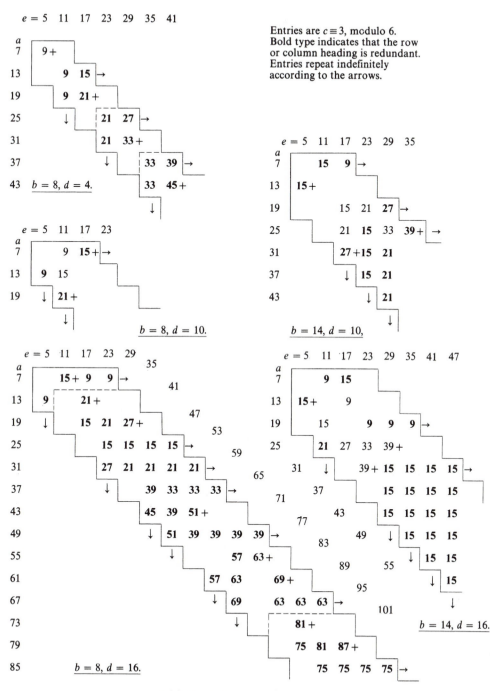

Entries are $c \equiv 3$, modulo 6.
Bold type indicates that the row
or column heading is redundant.
Entries repeat indefinitely
according to the arrows.

Table 7. $\mathcal{P}$-Positions Containing 6.

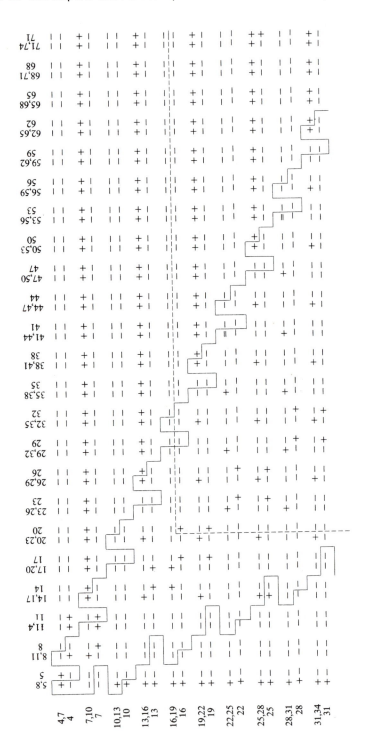

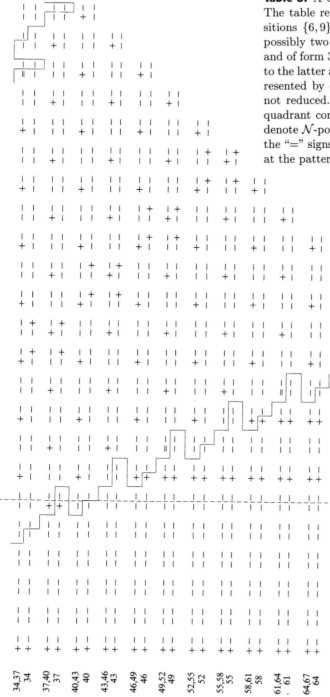

Table 8. A Complete Discussion of {6,9}.

The table represents a complete discussion of positions {6,9}. Reduced positions contain one, or possibly two neighboring numbers of form $3k+1$, and of form $3k+2$; pairs of rows and columns refer to the latter and former possibilities. Positions represented by cells outside the crenellated lines are not reduced. The pattern within the rectangular quadrant continues indefinitely. The minus sisgns denote $\mathcal{N}$-positions, the plus signs $\mathcal{P}$-positions and the "=" signs $\mathcal{N}$-positions which, at a casual glance at the pattern, might be mistaken for $\mathcal{P}$-positions.

For example, $\mathcal{G}(2, 2n+3) = n(n \geq 0)$, so $\mathcal{G}(2) = \omega$. On the other hand, $\mathcal{G}(3, 3n+1, 3n+2) = 1$ $(n \geq 1)$, so $\mathcal{G}(3) = 1$. Here are some other nim-values:

$k =$	4	5	6	7	8	9	10	11	13	14	15	16	17	19	20	22	23	25	26	28	29	31	32
$\mathcal{G}(3,k) =$	2	3	1	4	6	1	7	8	9	11	1	12	14	15	15	17	19	20	21	23	24	26	27
$\mathcal{G}(4,k) =$	?	3	0	5	?	8	1	9	14	4	15	?	16	19	?								

$$\mathcal{G}(4,6,4n-1,4n+1) = 1 \qquad \mathcal{G}(4,6,4n+1,4n+3) = 0 \qquad (n \geq 1)$$

$$\mathcal{G}(5,6) = 7, \quad \mathcal{G}(5,7) = 8, \quad \mathcal{G}(5,8) = 10, \qquad \mathcal{G}(6,7) = 9, \qquad \mathcal{G}(3,3n-1,3n+1) = 5$$
$$(n \geq 6),$$

$$\mathcal{G}(3,9n-8,9n-4) = \mathcal{G}(3,9n+2,9n+7) = \mathcal{G}(3,9n+8,9n+13) = 10 \qquad (n \geq 3).$$

A Few Final Questions

Is there any effective technique for computing the outcome and *all* good replies for the *general* position?

If the game is played "between intelligent players", is the first person to make the game bounded the loser?

Is there a winning strategy of bounded length?

Is there an $\mathcal{N}$-position with $g > 1$ for which all good replies lead to positions with $g = 1$?

Is $\mathcal{G}(4) = \omega + 1$? or is $\mathcal{G}(4) = 6$, say?

References and Further Reading

John D. Beasley, *The Mathematics of Games*, Oxford Univ. Press, Oxford, UK, 1989, chap. 10

Morton Davis, Infinite games of perfect information, *Ann. of Math. Studies, Princeton*, **52**(1963) 85–101.

Martin Gardner, Mathematical Games, *Sci. Amer.* **228** Jan. 111-113, Feb. 109, May 106–107, for David Gale's Chomp.

Richard K. Guy, Twenty questions concerning Conway's Sylver Coinage, *Amer. Math. Monthly*, **83** (1976) 634–637.

Scott Huddleston and Jerry Shurman, Transfinite Chomp, in Richard Nowakowski (ed.) *More Games of No Chance*, (Berkeley CA 2000) *Math. Sci. Res. Inst. Publ.*, **42** (2002) Cambridge Univ. Press, Cambridge, UK, 183–212.

Michael O. Rabin, Effective computability of winning strategies, *Ann. of Math. Studies*, Princeton, **39**(1957) 147–157: M.R.20#263.

Fred. Schuh, The game of divisions, *Nieuw Tijdschrift voor Wiskunde*, **39**(1952) 299–304.

George Sicherman, Theory and practice of Sylver Coinage, *Integers*, **2**(2002) G2 (electronic): updates the information on Sylver Coinage.

J. J. Sylvester, Math. Quest. *Educ. Times*, **41**(1884) 21.

Sun Xin-Yu, Improvements on Chomp, *Integers: Electr. J. Combin. Number Theory*, **2** (2002) G1; MR 2003e:05015 http://www.integers-ejcnt.org/vol2.html

[This extends Zeilberger's article by calculating $\mathcal{P}$-positions with an arbitrary number of rows versus three rows. It also presents a formula similar to ours in Fig. 12 by allowing the second column to have three pieces.]

Doron Zeilberger, Three-rowed Chomp, *Adv. in Appl. Math.*, **26** (2001) 168–179; MR 2001k:91009.

-19-

The King and the Consumer

For fools rush in where angels fear to tread.
Alexander Pope, *Essay on Criticism.*

... because your adversary the devil, as a roaring lion,
walketh about, seeking whom he may devour.
1 Peter 5:8

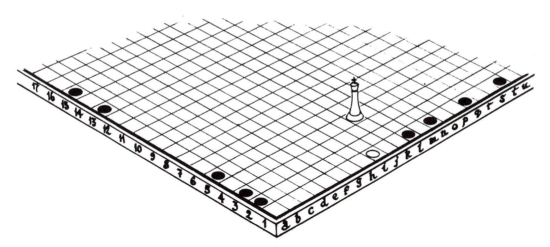

Figure 1. Chas. Plays Geo.

Chessgo, Kinggo and Dukego

These games are played on some i by j board. One player, Chas., plays Chess with a lone chess piece which might be a King, or a Knight, or a Duke, or a Ferz, whose moves are shown in Fig. 2. The variants of Chessgo are named for various real and Fairy Chess pieces, Kinggo for a King, etc. Only Kinggo and Dukego will be considered in any detail here.

641

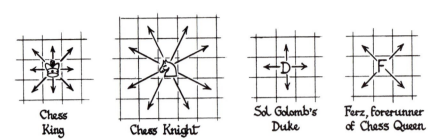

Figure 2. Various Chesspersons.

Chas.'s opponent, Geo., has a number of black (blocking) Go stones and a number of white (wandering) ones. The game starts with the chess piece on a specified square of the otherwise empty board. At each turn the chess piece moves to any legitimate empty square and Geo. then does one of the following:

(a) puts a new Go stone (of either color) on any empty square,

(b) moves a wandering (white) stone already on the board to any other empty square,

(c) passes.

If the chess piece reaches any square on the edge of the board, Chas. wins. If Geo. succeeds in surrounding the chess piece so that it has no legal moves, he wins. A game that continues forever is declared a draw.

Quadraphage

This is the special case, invented by R. Epstein, where there are no wandering stones and enough blocking ones to cover the whole board. The title of this chapter refers to the case of Quadraphage in which the chessperson is the King. In Epstein's language, Geo. is a square-eater (graeco-latin *tesseravore*, latino-greek *quadraphage*). Because Geo. eats a square at every turn, this game ends after at most $ij-1$ turns on an i by j board. The starting position for the chessperson is conventionally the middle of the board, or as near as possible if i or j is even.

Since having the first move is never a disadvantage, a strategy-copying argument shows that there are only three possible outcomes for a well-played Quadraphage game from a given starting position on a finite board. Either Geo. wins (even if Chas. moves first) or Chas. wins (even if Geo. moves first) or the first player to move wins. A **fair position** is one in which the first player to move can win.

We'll show that the fair starting positions for the Duke on a quarter-infinite board are all of the squares on the third rank or file, *except* those that are also on the first or second file or rank. We'll also show that the fair starting positions for the King on this board are all of the squares on the ninth rank or file, except those which are also on a lower file or rank. Finally we'll show that the square board which is fair (from the conventional starting position) for a Duke is the ordinary 8 by 8 chessboard, and we assert that the only fair and square boards for a King are 33 by 33 and 34 by 34. On boards smaller than these, Chas. should win even if Geo. starts first and the reverse should happen on larger boards.

The Angel and the Square-Eater

The game of Chessgo is not well understood and it's very difficult to exhibit explicit winning strategies for Chas. even on modest sized boards. For example it seems very likely indeed that the Knight can draw on an infinite board although this seems extremely difficult to prove.

Indeed it's never been shown that there is *any* generalized chess piece that can draw on the infinite board. This suggests the following problem. An **angel** (of power 1000) is a chessperson who can fly in one move to any empty square that could be reached by a thousand King moves. Angels, of course, have wings, so it won't matter if some of the intervening squares have been eaten.

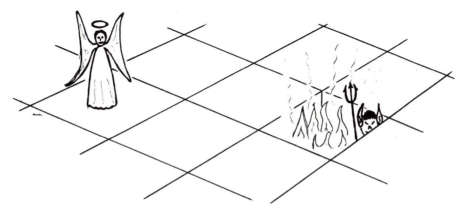

Figure 3. The Angel and the Square-Eater.

We'll say the angel wins by continuing forever (i.e. drawing the game of Quadraphage) against a square-eating **devil** (who can devour *any* square of the board, no matter how far away it is from his previous moves). The devil, of course, wins if he can surround the angel with a sulphurous moat, a thousand squares wide, of eaten squares. Can you give an explicit strategy that's *guaranteed* to win for the angel?

If the devil adopts certain cunning tactics worked out for him by Andreas Blass and John Conway, then infinitely often the angel will find itself decreasing its distance from the centre by arbitrarily large amounts. Although the angel never seems to be in any real danger, its path must also contain arbitrarily convoluted spirals.

Strategy and Tactics

In both Dukego and Kinggo it's possible to distinguish between strategic moves and tactical ones. In either game Geo. wins, on large enough boards, by first playing a few *strategic* stones on squares far away from the chess piece. When the chess piece gets closer to the edge of the board, Geo. switches to *tactical* moves fairly close to him. Whenever the chess piece is driven away from the edge towards the centre of the board, Geo. reverts to strategic moves.

Dukego

Dukego is much simpler than Kinggo and so we consider it first. You might like to try playing it yourself before reading this section. The optimal strategies we present here were first discovered for square boards by Solomon Golomb and for arbitrary rectangular boards by Greg Martin. We consider various infinite boards first.

On an infinite half-plane the Duke can win only if he can get to the edge at his first move. In any other situation Geo. can draw by playing directly between the Duke and the edge. In fact Geo. needs only one white (wandering) stone.

On an infinite strip of width i with $i \leq 4$ the Duke, moving first, can win immediately. If $i \geq 4$ and Geo. moves first he can draw by playing between the Duke and the nearest edge and again needs only one stone, if it's a wandering one.

On an infinite quarter-plane the Duke, moving first, can win if he starts within a three squares wide border. His initial move attacks the edge and Geo. has no choice but to move directly between Duke and edge. The Duke then charges towards the corner. At each move Geo. is forced to play between the Duke and the edge and eventually the Duke wins by reaching one of the two squares next to the corner.

If Geo. moves first against the Duke on the third rank or file of an infinite quarter plane, he can draw using just one blocking stone and one wandering one. He first puts his blocking stone at the strategic position diagonally next to the corner (Fig. 4). This blocks the only square from which the Duke might attack two boundary squares at once. Whenever the Duke moves onto a lower case letter, Geo. puts his wandering stone on the corresponding capital letter.

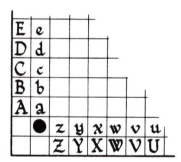

Figure 4. Geo. Beats the Duke on a Quarter-Infinite Board.

On the 8 by 8 board Geo. can draw using only three wandering stones (Fig. 5a). He always arranges to have his stones on the capital letters corresponding to the small letters covered by the Duke. Since the combinations of small letters on any two contiguous squares never differ by more than one letter, this is always possible.

We once thought that the Duke could win on a $7 \times j$ board, even if Geo. starts, but Greg Martin showed that this is incorrect. His clever strategy is shown in Fig. 5b, with the Duke starting on the center square labelled b: Geo. places his first stone at B.

Martin also devised the strategy shown in Fig 5c, by which Geo., moving first, can win on the 6×9 board. Here the Duke starts on the centre square labelled cf and Geo. places his

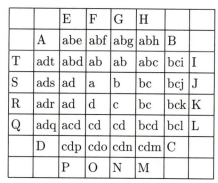

		E	F	G	H		
	A	abe	abf	abg	abh	B	
T	adt	abd	ab	ab	abc	bci	I
S	ads	ad	a	b	bc	bcj	J
R	adr	ad	d	c	bc	bck	K
Q	adq	acd	cd	cd	bcd	bcl	L
	D	cdp	cdo	cdn	cdm	C	
		P	O	N	M		

Figure 5a. Geo. Beats the Duke on an Ordinary Chessboard.

		F	G	H	I		
	A	abf	abg	abh	abi	B	
R	aer	be	ab	ab	abc	bcj	J
Q	aeq	ae	be	b	bc	bck	K
P	aep	ade	de	bd	bcd	bcl	L
	E	deo	den	D	cdm	C	
		O	N		M		

Figure 5b. Geo. wins on 7 × 8 board.

		H		I		J		
	A	abh	B	bci	C	cdj	D	
R	agr	abg	bg	bc	ce	cde	dek	K
Q	agq	afg	fg	cf	ef	def	del	L
	G	fgp	fgo	F	efn	efm	E	
		P	O		N	M		

Figure 5c. Geo. wins on 6 × 9 board.

first stone directly beneath it at F. If the Duke moves northward, then Geo. simply reflects the board about its horizontal centerline, and moves his stone from the old F to its reflexion, which is immediately north of the Duke. As long as the Duke moves only north and south, Geo. can continue to respond by moving his single stone directly in front of him. Eventually the Duke will make a move east or west and then Geo. can play according to Fig 5c or its reflexion.

The Geo. strategy shown on all three boards of Fig. 5 uses three wandering stones. But Geo. can also draw using only two wandering stones and two blocking stones. On the 8 × 8 board this is accomplished simply by placing the blocking stones on A and C. This works because every square with three letters in Fig. 5a includes at least one of a and c.

Greg Martin has shown that two wandering stones, or just one wandering stone and four blocking stones, are also sufficient for Geo. to draw on the the 7 × 8 and 6 × 9 boards. It is much harder to find out how many blocking stones Geo. needs when he has no wandering stones.

Table 1 summarizes the fair starting positions in Dukego. Martin's paper includes a concise proof that the Duke, going first, can win on 8 × 8 or on 6 × j for any value of j.

Size of Board	Starting Position	Least Number of Stones Giving Geo. at least a Draw, Moving First
$4 \times \infty$	centre	1 wandering
quarter-infinite	3rd rank or file excluding 1st or 2nd file or rank	1 wandering
8×8 7×8 $6 \times j, j \geq 9$		3 wandering, *or* 2 wandering, 2 blocking, *or* 1 wandering, 4 blocking, *or* ? blocking

Table 1. Fair Boards for Dukego.

The Game of Kinggo

The remaining sections of this chapter are devoted to Kinggo.

The Edge Attack

Figure 6 shows how the King can force his way to a nearby edge of the board if this is inadequately defended. The solid line indicates the edge of the board and the dot shows the present position of the King. We suppose that the lettered squares are empty; Go stones may occupy any or all of the other squares. In each case Geo. is to move.

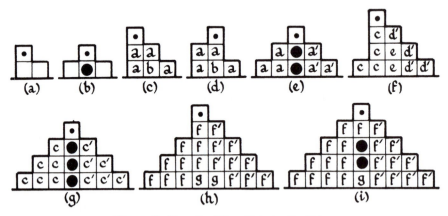

Figure 6. How the King Gets to the Edge.

If Geo. moves outside the region shown, the King simply advances towards the edge, while if Geo. moves on to a letter x or x' (x = a,b,...,g) the King makes a move which results in a case of Fig. 6(x) or its reflexion. If, for example, Geo. puts a stone in the lower right corner (labelled d') of Fig. 6(f), the King moves downwards and achieves a reflexion of Fig. 6(d).

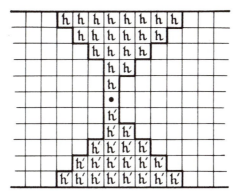

Figure 7. How the King Wins on an Infinite Strip of Width Eleven.

Now a glance at Fig. 7 and Fig. 6(h) shows that:

> the King can win
> on an infinite strip
> of width at most 11,
> even if Geo. goes first.

The Edge Defence

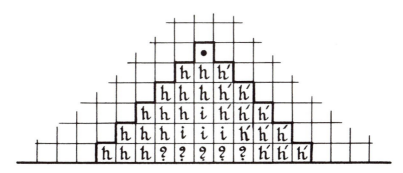

Figure 8. How Geo. Guards the Edge Against the King.

Figure 8, which again refers back to Fig. 6, shows that there are only five possible moves (?, ?, ?, ?, ?) that give Geo. any chance of stopping the King approaching from the sixth rank of an empty board. Figure 9(k) shows how Geo. can successfully defend the edge with any of these five moves. The King may move from any of the shaded squares. If he remains on such a square Geo. passes. When the King moves on to a letter x or x' (x = j,k,l,...,q) Geo. can move into a case of Fig. 9(x) or its reflexion (as he did in Fig. 6). Note that the proof of

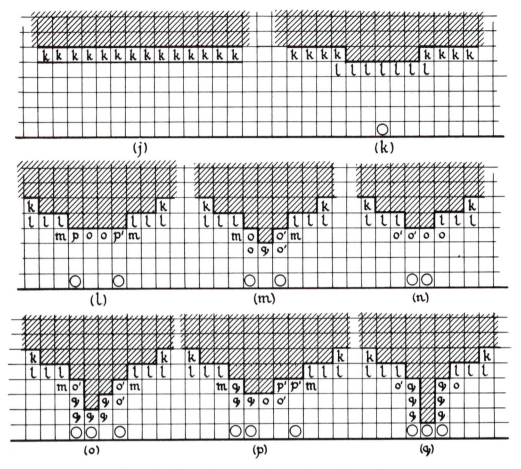

Figure 9. Three Wandering Stones Ward Off the King.

each of Figs. 9(j) to 9(q) depends on the others, because the King can nip from one of these positions to another in ingenious ways.

Since none of these positions has more than three stones, Geo. can defend the edge with only three wandering stones.

A Memoryless Edge Defence

Figure 10, which uses the same conventions as Fig. 5, shows another way that Geo. can stop the King approaching from the sixth rank of an empty board using just three wandering stones. Unlike Fig. 9, this is memoryless in the sense that the positions of these stones depend only on the position of the King and not on how he got there.

In later sections of this chapter Geo. will want to patch together several copies of this defence (Figs. 11 and 12). Figure 11 shows how it may be joined to its left-right mirror

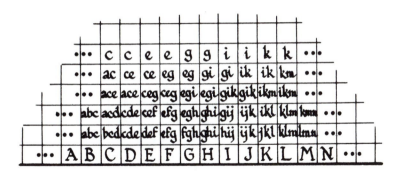

Figure 10. A Memoryless Kinggo Edge Defence.

image, and Fig. 12 shows how to change its phase by one square. Many other memoryless edge defences can be obtained by joining various combinations of Figs. 10, 11, 12 and their translates and reflexions.

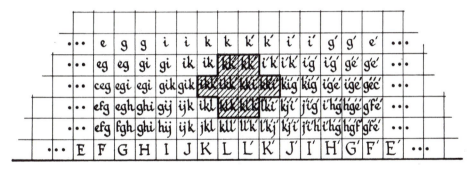

Figure 11. Wedding Figure 10 with Its Reflexion.

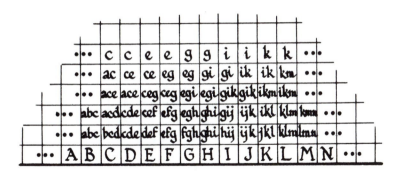

Figure 12. Getting Figure 10 One Square Out of Step with Itself.

Some results for infinite strips follow immediately from Figs. 9, 10, 11, 12:

> On an infinite strip
> of width at least 12,
> Geo., moving first, can draw
> with just 3 wandering stones.
>
> ———
>
> On an infinite strip
> of width at least 13,
> he can draw even if
> the King moves first.

If the King advances towards the edge he will be stopped at Fig. 9(q) and if the King refuses to attack the edge, Geo. can still obtain 3 consecutive stones as in that figure.

The Edge-Corner Attack

On an infinite strip of width 13, the King can't *win*, but *can* force his way to the second rank, as in Fig. 9(q). He may then charge along this rank in either direction, forcing Geo. to accompany him. Even if Geo. has a large supply of stones he can do no more than build up a solid wall along the first rank and on a finite board the edge-charging King will eventually reach a corner.

We now claim that for an adequate defence, Geo. must have at least three strategic stones stationed somewhere between the edge-charging King and the corner. All of these three stones must be positioned somewhere in the first five ranks. The proof of this follows from Fig. 13, which lists the appropriate moves for the King against all positions not satisfying these conditions. There are squares with one of the first seven capital letters and infinitely many squares with no letters at all. At Geo.'s move he has at most 3 stones in the figure. The line

$$\text{``Versus } A^0, \text{ go to } A, \text{ position } \text{—''}$$

means that if none (superscript 0) of the 3 stones are on A, then the King moves to A and wins at once. The line

$$\text{``Versus } A^1 B^0 C^0 D^{\leq 1} E^{\leq 1}, \text{ go to } C, \text{ position } a\text{''}$$

means that if Geo. has one stone on A, none on B or C, and at most one (superscript ≤ 1) on each of D and E, the King should move to C and obtain a translate of the position shown in Fig. 13(a).

Since in every case the King counters Geo.'s moves to any of Figs. 13(a) to 13(g) by a move resulting in another of these figures, Geo. can never force the King above the fifth rank or prevent him from continuing the edge-corner attack, although he can keep him moving to and fro among these seven figures.

On the other hand, almost all combinations of three strategic stones along the first rank of the board *will* suffice for Geo. to stop the edge-corner attack. Figure 14 shows the only exceptions.

	Versus	Go to	New Position
(a) At most two stones. Not both on A and D, Not both on two of A, C, F.	A^0	A	
	$A^1 B^0 C^0 D^{\leq 1} E^{\leq 1}$	C	a
	$A^1 B^0 C^0 D^0 E^2$	F	c
	$A^1 B^0 C^0 D^2$	prohibited	
	$A^1 B^0 C^1 F^0$	F	b
	$A^1 B^0 C^1 F^1$	prohibited	
	$A^1 B^1 F^0$	F	b
	$A^1 B^1 F^1$	C	a
(b) At most one stone.	$A^0 B^{\leq 1} C^{\leq 1}$	A	a
	$A^0 C^2$	B	c
	$A^0 B^2$	D	d
	$A^1 B^0$	B	b
	$A^1 B^1$	D	d
(c) Stones on two of B, C and D. None elsewhere.	A^0	A	d
	$A^1 B^0 C^1 D^1$	E	$-$
	$A^1 B^1 C^0 D^1$	E	a
	$A^1 B^1 C^1 D^0$	E	a
(d) No stones.	A^0	A	f
	A^1	B	e
(e) No stones.	A^0	A	f
	A^1	B	e
(f) At most one stone.	A^0	A	g
	$A^1 B^0$	B	f
	$A^1 B^1$	C	e
(g) At most two stones.	$A^{\leq 1}$	A	$-$
	$A^2 B^0$	B	b
	$A^2 B^1$	C	d

Figure 13. The Edge-Corner Attack.

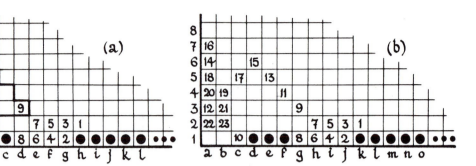

Figure 14. Triplets Which Fail to Stop the Edge-Charging King.

In Fig. 14(a) Geo. has three strategic stones at a1, b1, c1, as well as his tactical stones, h1, i1, ..., defending the edge near the King. The King moves to 1; Geo. is forced to put a stone at 2; the King moves to 3; and so on. The King's move to 9 guarantees him a win by Fig. 6(f) (reflected).

Figure 14(b) uses a similar notation to show how the King wins if Geo.'s strategic stones are at d1, e1, f1.

Strategic and Tactical Stones

Since Geo. can stop an edge corner attack with just three extra stones, and most combinations of three stones suffice, it's convenient to call his three most distant stones in the first five ranks along any edge of the board **strategic** stones; his other stones are **tactical** ones. The tactical stones try to stop the King winning along the side and the strategic ones then prevent him from winning the ensuing edge-corner attack.

Let's consider for instance the game on an infinite strip of width 23 (Figure 15). The King starts at 1; Geo. puts a stone at 2; the King moves to 3; Geo. puts a stone at 4; and so on. The crucial position arises when Geo. puts a stone at 16. Where should the King move now? Although various moves look plausible, only one succeeds!

You must distinguish between strategy and tactics if you're to find the right move. The stones 4, 6 and 8 defend the right flank, so 10, 12 and 16 are needed to defend the left one. Since the stone at 16 is required for *strategic* purposes it is *tactically* worthless.

So the King pretends that 16 is empty and moves to 17, which would give him a tactical victory via Fig. 6(g)! Any other King move would lose to a defence at α or β.

Of course, since 16 *isn't* empty, the game *won't* end on the lower edge, for Geo. can stop the edge attack, but only by using the stone at 16. Eventually the King would have an opportunity to move to 16 if it were vacant. Instead of doing this he embarks on an unstoppable edge-corner attack, running along the second rank towards the left. Geo. can eventually use his stones 10 and 12 to divert the King into various positions of Fig. 13 but can't halt the edge-corner assault.

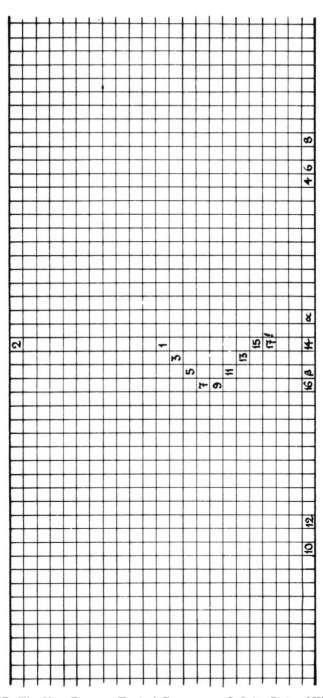

Figure 15. The King Draws a Typical Game on an Infinite Strip of Width 23.

This sort of argument shows that:

> if Geo. is required to place
> his first 10 stones on the
> top and bottom ranks, the
> King can draw on an
> infinite strip of width 23,
> even when Geo. moves first

We believe that this remains true when we remove the constraints on Geo.'s initial moves, since it seems very unlikely that he gains any advantage by putting his stones nearer to the middle. Although such moves seem futile, we haven't managed to exhibit a precise strategy by which the King can refute them.

Corner Tactics

Figure 16 shows how Geo. defends the corner against an attack from either edge using three consecutive blocking stones and three wandering ones. The edges can be continued using Fig. 10 to give a strategy for Geo. on a quarter-infinite board.

Although it defends the corner from attack along either edge, this provides only a weak defence against a direct attack towards the corner along the diagonal. It defends against a King on the tenth rank and tenth file of an empty board *only* when Geo. moves first. Since Geo. must first enter his three strategic stones, if the King moves first he will arrive at the sixth rank and file before Geo. has put any wandering stone on the board, and Fig. 16 now requires Geo. to put wandering stones on both F and X. In fact Fig. 17 (which should be used in conjunction with Fig. 6) shows that the King can now win against *any* strategy for Geo., even if there are stones on all the indicated squares of Fig. 17(a). This figure depends on Figs. 17(b) to 17(e) whose proofs are left to the reader. Figure 17 shows that Geo.'s only hope of defending the corner against a diagonally attacking King, starting from the tenth rank and file, requires that his first three stones be placed elsewhere. One promising possibility uses squares a2, a3, a5 along one edge, when Geo.'s major problem is to find an appropriate continuation when the King arrives on the sixth rank and file. Figure 18(a) shows that there is only one possibility (indicated by ?). The proof of Fig. 18 depends on Fig. 17 and Fig. 6, but we again leave some of these proofs to the reader.

So there's only one move with which Geo. can successfully defend Fig. 18(a)! His complete strategy appears in Fig. 19 with the edges extended by Fig. 10. With the three blocking stones positioned as shown, he defends the corner against attacks along edges or diagonal.

Combining Figs. 13 and 19 we have:

> the fair starting positions
> on the quarter-infinite board
> are those on the ninth rank or file,
> excluding lesser files or ranks.

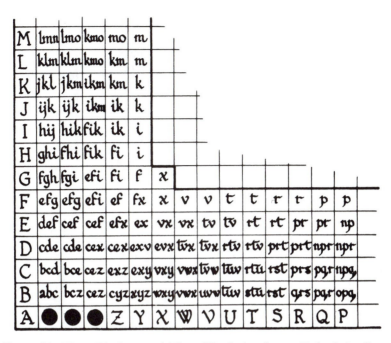

Figure 16. Three Blocking and Three Wandering Stones Defend the Corner.

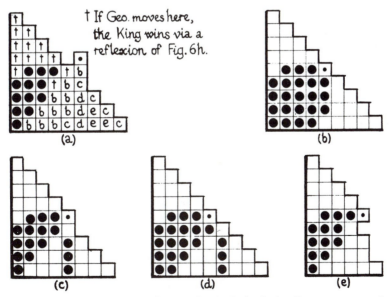

Figure 17. Three Consecutive Blocking Stones Won't Defend the Corner against the King on the Sixth Rank and File.

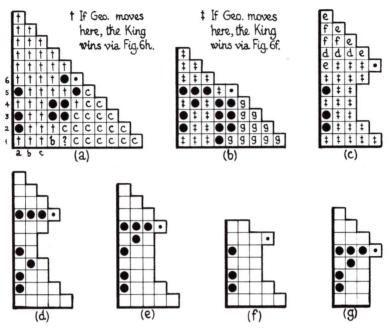

Figure 18. With Three Well Placed Stones and the Initial Move, Geo. Defends the Corner against the Diagonal Attack.

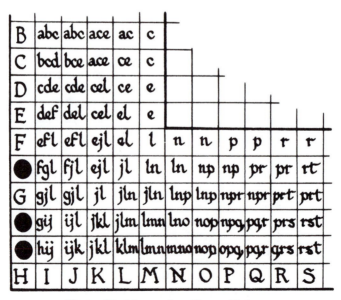

Figure 19. Memoryless Corner Defence.

Geo., moving first, can defend any such position with just three blocking stones and three wandering ones according to Fig. 19. But if the King moves first, he attacks the nearest edge, ignoring the three further stones between him and the corner as in the sample game of Fig. 15. Since Fig. 13 shows that Geo. needs three strategic stones to defend the corner, the King can *either* win on the edge *or* divert to a winning edge-corner assault as in Fig. 13.

Defence on Large Square Boards

We've seen that Geo. can defend a corner with three blocking stones and three wandering ones, so he can defend a large enough square board with twelve blocking stones (three in each corner) and three wandering ones. He first puts the twelve blocking stones in their permanent places. If the board is 35 × 35 or bigger, the King begins at least 18 squares from any edge, so he's still at least 6 squares away from the edge after Geo. has placed his 12-strategic stones. So,

> Geo., moving first,
> can win on a square board
> of size 35 × 35 or larger.

THE 33 × 33 BOARD

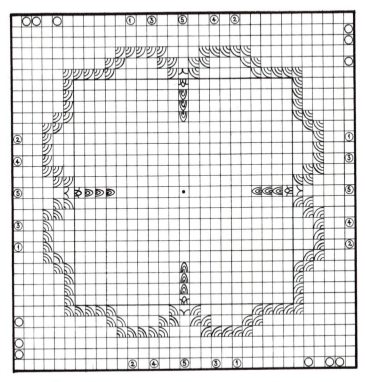

Figure 20. The Centred King on a 33 × 33 Board.

The 33 × 33 Board

We'll now show you a more intricate defence which allows Geo., moving first, to survive on a 33 × 33 board with just 12 (wandering) stones. The details are in Figs. 20 to 26.

The Centred King

So long as the King stays in the central region of Fig. 20, Geo. puts stones on certain strategic squares, marked with circles on the perimeter of the board. There are 32 of these, 3 near each corner and 5 on each edge. Geo. puts the first four stones one on each edge, and the distribution of his stones after the King has made four or more moves is shown in Fig. 21, a close-up of part of Fig. 20 (the four quarters of the board are congruent). Most of the squares in the central region are divided into nine subsquares, the central one of which is always empty. The other eight subsquares tell Geo. how many stones he should have in each corresponding area. For example, if the King moves to a square marked

then Geo. moves so that he has three stones on the left edge, one near the bottom left corner and four on the bottom edge. The order in which Geo. puts his stones in the three squares near the corner doesn't matter, but of the five strategic squares on each edge, it's the middle one that must be occupied last. A reasonable order is indicated by the numbers 1,2,3,4,5 in the circles in Figs. 20 and 21.

A few squares on the main diagonals of Fig. 21 contain arrows:

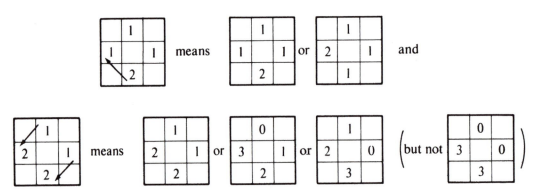

Geo. can use any of these alternatives as a satisfactory defence.

Leaving the Central Region

If the King leaves the central region of Fig. 20 via a square marked as in Fig. 22(a), we'll say that he's **cornered** in the lower left of the board, and then Geo. will keep him inside the region

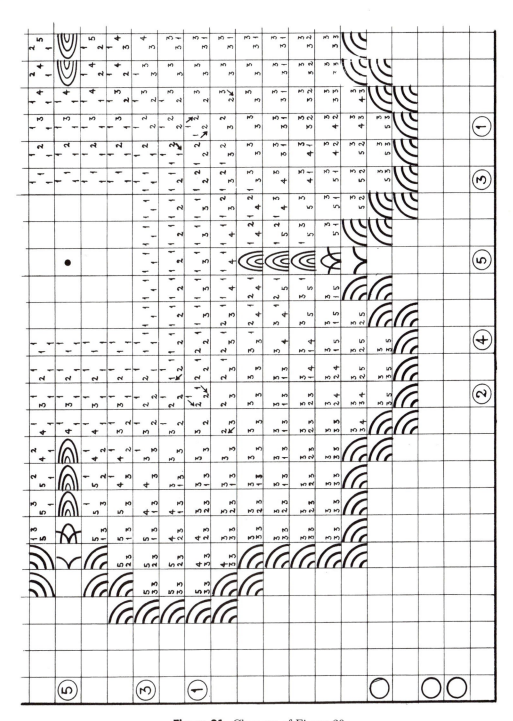

Figure 21. Close-up of Figure 20.

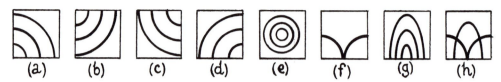

Figure 22. Key to Markings in Figures 20, 21, 23, 24, 25 (see text).

shown in Fig. 23 by making tactical moves that prevent the King from reaching a shaded square, so that he can only "re-centre" himself by moving to a square marked as in Fig. 22(e).

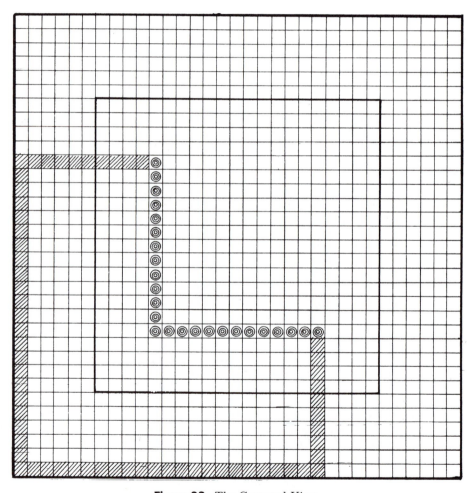

Figure 23. The Cornered King.

If the King moves to squares marked as in Fig. 22(b), (c) or (d) he is correspondingly cornered in the upper left, upper right or lower right of the board. If the King moves to a square labelled as in Fig. 22(f) he is cornered in the lower left or lower right of the board, depending on the direction he came from. If he moves to a label like Fig. 22(g) he is **sidelined** (see later) and when he moves to one like Fig. 22(h) he is either sidelined or cornered, again depending on which way he came; if diagonally, he's sidelined; if horizontally then he'll be pushed back to the corner whence he came.

The Cornered King

Figure 24, a close-up of Fig. 23, reveals the tactical details that Geo. uses to keep the King cornered with just three wandering stones and nine **static** ones (semi-stationary, both strategic and tactical). Of course, when the King first becomes cornered by moving to a square marked as in Fig. 22(a), Geo. may not have his nine static stones in the exact places shown in Fig. 24, but he will have three stones between the King and the lower left corner, and three on the bottom edge and three on the left edge. Geo. uses the stones already on the boundary as substitutes for any stones missing from Fig. 24. When the tactics call for placing a stone on a square already occupied, Geo. places a stone on an unoccupied circle in Fig. 24.

Suppose, for example, that the King leaves the central area of Fig. 20 by moving to square $k4$ (see Fig. 1). He must have come from $l5$, marked so there are already three stones on the

<table>
<tr><td></td><td></td><td></td></tr>
<tr><td>3</td><td></td><td></td></tr>
<tr><td>3</td><td>5</td><td></td></tr>
</table>

left-edge, three as indicated near the lower left corner and five in the squares 2, 4, 5 and those next to Z and A and between them ("3" and "1"). The King is now on a square labelled "$s24$" so Geo. puts his last stone (the white one in Fig. 1) on S and continues to follow Fig. 24 with the stone on "5" substituting for the missing one between Z and A.

The right arrow in certain squares near the top right of Fig. 24 means that the third genuinely wandering (non-static) stone belongs on a strategic square on the right edge.

The Sidelined King

If the King leaves the central area of Fig. 20 by moving onto a square marked as in Fig. 22(g) he is sidelined as in Fig. 25. Geo. tactically keeps the King off the shaded squares and the King can only re-centre himself by moving onto a square marked as in Fig. 22(e).

The notation in Fig. 26 (a close-up of Fig. 25) is as in Figs. 21 and 24, but we now have some squares.

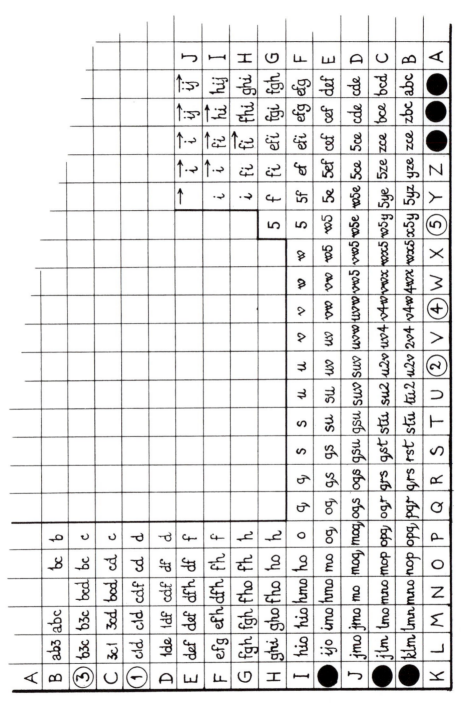

Figure 24. Close-up of Figure 23.

These advise Geo. to have one stone on each of the left and right edges (shown in Fig. 25 in their lowest and highest positions) and *six* on the bottom edge, not only the usual five but one on J or Q as well. [Assume in Fig. 26 that Geo. has 6 static stones on $1, 2, 3, 4, J, Q$, and that one of the last two is substituting for 5.]

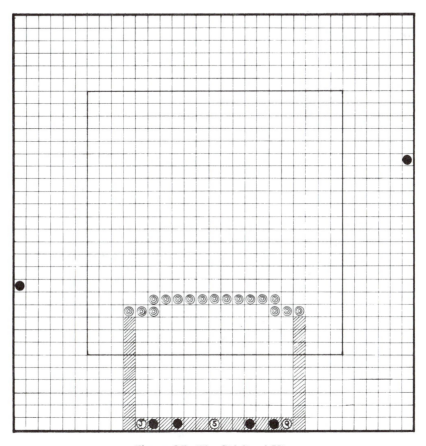

Figure 25. The Sidelined King.

How Chas. Can Win on a 34 × 34 Board

Geo., going first, can survive in Kinggo with 12 wandering stones on a 33×33 board, so he can certainly survive on a 35×35 or larger board, even if the King goes first.

However it seems that the King can win on a 34×34 board if he moves first. Here's how he does it. His first three moves diagonally attack the nearest corner. He then turns left or right and attacks the adjacent corner in the half of the board where Geo. has at most one stone. After 9 more moves he is at $l6$, say, and Geo. has been unable to get 9 useful stones on the board. If the corner is adequately defended (with 3 stones), then one flank or the other is weak and a carefully executed edge-corner attack eventually leads to victory.

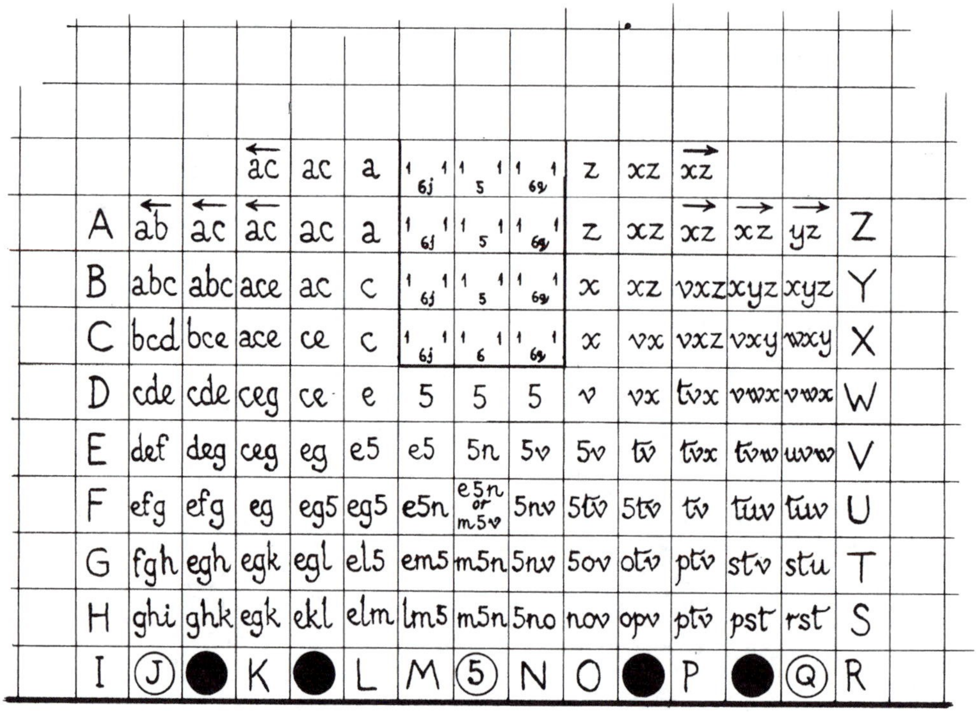

	1	2	3	4	5	6	7	8	9	10	11	12	13	
			←ac	ac	a	1 6j 1	1 5 1	1 6y 1	z	xz	xz			
A	←ab	←ac	←ac	ac	a	1 6j 1	1 5 1	1 6y 1	z	xz	→xz	→xz	→yz	Z
B	abc	abc	ace	ac	c	1 6j 1	1 5 1	1 6y 1	x	xz	vxz	xyz	xyz	Y
C	bcd	bce	ace	ce	c	1 6j 1	1 6 1	1 6y 1	x	vx	vxz	vxy	wxy	X
D	cde	cde	ceg	ce	e	5	5	5	v	vx	tvx	vwx	vwx	W
E	def	deg	ceg	eg	e5	e5	5n	5v	5v	tv	tvx	tvw	uvw	V
F	efg	efg	eg	eg5	eg5	e5n	e5n or m5v	5nv	5tv	5tv	tv	tuv	tuv	U
G	fgh	egh	egk	egl	el5	em5	m5n	5nv	5ov	otv	ptv	stv	stu	T
H	ghi	ghk	egk	ekl	elm	lm5	m5n	5no	nov	opv	ptv	pst	rst	S
I	(J)	●	K	●	L	M	(5)	N	O	●	P	●	(Q)	R

Figure 26. Close-up of Figure 25.

Unfortunately we haven't been able to formalize these remarks into a strategy for Chas. that's even as explicit as Geo.'s 33×33 one.

Rectangular Boards

Geo. can't beat the King on the infinite strip of width 23, even if he moves first and has an unlimited supply of stones. However, if he moves first on a 24 by n board, he can win for sufficiently large n. The minimum value of n seems to be about 63. The King is immediately sidelined along whichever long edge he's nearest to. The King can circumvent the pseudo-corners (I and R in Fig. 26) of Geo.'s sideline defence, but only by moving back to squares about midway between the two long edges. Geo. can then defend a second pseudocorner between the real corner and the first one. By the time the King reaches the corner, Geo. has prepared defences of both corners along a short edge of the board and a pair of opposite pesudocorners somewhere between the King and the unattacked short edge.

For each value of i, $24 \leq i \leq 37$, there appears to be a range of values of j for which the i by j board is a fair battleground for a Quadraphage game against the Chess King. We believe that the 32×33 board is fair and that Geo., moving first, wins with a strategy similar to that we gave for the 33×33 board. We leave the problem of determining the dimensions of all fair Quadraphage boards as a challenge for Omar.

Extras

Many-Dimensional Angels

Many-dimensional angels can escape from the corresponding hypercube-eaters. This has been proved by Tom Körner who thinks that his proof could conceivably be adapted to the two-dimensional game. Don't write to us with your solution to the Angel problem unless you've taken account of the remarks on p. 643.

Games of Encirclement

The games of this Chapter, and of the next two, are ones of *encirclement* or *escape*. There are many games, going a long way back in history, in which this idea is combined with varying kinds of *capture*. Here are a few examples.

Wolves-and-Sheep

There are several games played on Solitaire-like boards (Chapter 23). In **Wolves-and-Sheep** (Fig. 27(a)) the shepherd has 20 sheep, which have first move. They move one place forward or sideways only, onto unoccupied places. The two wolves can move similarly but on any of the indicated lines and can capture in these directions by jumping as in checkers (draughts), including multiple captures. A wolf failing to make a possible capture may be removed by the shepherd, so the sheep may be used as decoys. The shepherd wins if he gets nine sheep into the *fold* (top 9 positions of board).

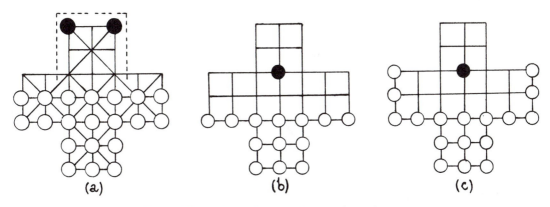

Figure 27. Wolves, Sheep and Other Animals.

The games shown in Figs. 27(b) and (c) are called Fox and Geese, although we use this name for a different game in the next chapter. They are similar to Wolves-and-Sheep, but there are no diagonal moves. The fox starts in any unoccupied position, and the geese try to crowd the fox into a corner; In Fig. 27(b) the 13 geese can move in any of the four orthogonal directions, but the 17 geese of Fig. 27(c) can't move backwards; they move like the sheep in Wolves and Sheep.

Hala-tafl (the Fox Game), and **Freystafl** are mentioned in the later Icelandic sagas. As in the Chapter 20 version of Fox and Geese, the more numerous animals win with correct play, but it's very easy to make mistakes!

Tablut

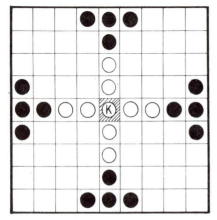

Figure 28. The Start of a Game of Tablut.   **Figure 29.** A Muscovite Captures Two Swedes!

Linnaeus, on his 1732 visit to Lapland, recorded a game played on a 9 × 9 board (Fig. 28) whose centre square, the **Konakis** or throne, may only be occupied by the Swedish King. He is protected by 8 blond Swedes and confronted by the 16 swarthy Muscovites. All the pieces move like the rook in Chess, any distance orthogonally. Capture of the King is by surrounding him, N, S, E and W by four Muscovites or by three Muscovites with the Konakis as the fourth square. Any other piece is removed by **custodian** capture, i.e. by placing two opposing pieces to the immediate N and S, or E and W of it. Figure 29 snows a Muscovite capturing two Swedes. A piece may move "into custody" without being captured. The aim of the Swedes is to get their King to the edge of the board.

Saxon Hnefatafl

Only a fragment of a board has been found; it is probable that the game was played using the 19 x 19 positions of a modern Go board. See R.C. Bell's excellent little book for a possible reconstruction from a tenth century English manuscript. The game was evidently like Tablut apart from the size of the board and the number and position of the pieces.

We finish this chapter with two Chess problems which also involve escape or encirclement.

King and Rook Versus King

Most beginning Chess players soon learn how to win this ending, so it's a surprise to find a couple of non-trivial problems which use just this material, albeit on a quarter-infinite board.

In Fig. 28, can White win? If so, in how few moves? Simon Norton says it's better to ask, "what is the smallest board (if any) that White can win on if Black is given a win if he walks off the North or East edges of the board?" Can Omar prove that it's 9×11?

Figure 29 shows Leo Moser's problem: can White win if he's allowed to make *only one move with the Rook*? If you find yourself frustrated by this, partition the squares in the first three columns into the four sets

> a1,a3,a5,...,c2,c4,c6,...
> b1,b3,b5,...
> a2,a4,a6,...,c1,c3,c5,...
> b2,b4,b6,....

Simon Norton's problem derives from an earlier Kriegspiel problem in which the white king and rook begin on adjacent locations near the corner of an infinite quadrant, and the black king starts at an arbitrary square within the quadrant. Tom Ferguson has outlined a mixed strategy for White that claims to eventually checkmate the black king with probability one.

Figure 29. Leo Moser's Problem.

Figure 28. Simon Norton's Problem.

References and Further Reading

Robert Charles Bell, *Board and Table Games from Many Civilizations*, Oxford University Press, London, 1969.

John H. Conway, The angel problem, in Richard Nowakowski (ed.) *Games of No Chance*, (Berkeley CA 1994) Math. Sci. Res. Inst. Publ., **29** (1996) Cambridge Univ. Press, Cambridge, UK, 3–12; MR 97m:90123.

Richard A. Epstein, *Theory of Gambling and Statistical Logic*, Academic Press, New York and London, 1967, p. 406.

Martin Gardner, Mathematical games: Cram, crosscram, and quadraphage: new games having elusive winning strategies, *Sci. Amer.* **230** #2 (Feb. 1974) 106–108.

C. Linnaeus, *Lachesis Lapponica*, London, 1811, p. ii, p. 55.

Greg Martin, Restoring fairness to Dukego, in Richard Nowakowski (ed.) *More Games of No Chance*, (Berkeley CA 2000) Math. Sci. Res. Inst. Publ., **42** (2002) Cambridge Univ. Press, Cambridge, UK, 79–87.

H. J. R. Murray, *A History of Board Games other than Chess*, Clarendon Press, Oxford, 1952.

David L. Silverman, *Your Move*, McGraw-Hill, 1971, p. 186.

-20-

Fox and Geese

While the one eludes, must the other pursue.
Robert Browning, *Life in a Love.*

The twelve good rules, the royal game of goose.
Oliver Goldsmith, *The Deserted Village*, I.232.

Figure 1. Playing a Game of Fox and Geese.

The game of Fox and Geese is played on an ordinary checkerboard between the *Fox*, who has just one black or red piece, and the *Geese*, who have four white ones. The players use squares of only one color (as in Checkers), and the Geese are initially placed in the squares marked O in Fig. 2. The Fox is usually placed at X in Fig. 2, but since the Geese seem to have the better chances, it is perhaps wiser to allow the Fox to choose his own starting square (provided this has the correct color), and then let the Geese have first move.

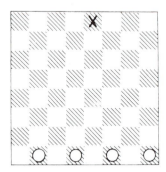

Figure 2. The Usual Starting Position.

The Geese move diagonally one place forward—like ordinary checkers they may not retreat. The Fox also moves diagonally one place, but like a King in Checkers, he may move in any one of the four diagonal directions. There is no taking or jumping. The Geese aim to trap the Fox so that he has no legal move, while conversely the Fox tries to break through the barrier of Geese so that he can stay alive indefinitely. We can therefore say simply that the first player unable to move is the loser, the usual normal play convention.

It is the general opinion that between expert players the Geese should win, but even against most moderately competent players a wily Fox can usually win a game every now and then, and if we let him choose his starting position, he should be able to defeat most novices for quite a long time. Perhaps those of our readers who have not met the game should take some time off to play a few games before reading further.

The question we shall ask and answer in this chapter is just how much of an advantage do the Geese have in this game? Perhaps we should first of all prove that the Geese really do have a winning strategy, even when the Fox is allowed the extra dispensation we suggest. In Fig. 3 we show the five types of position that our own favorite strategy relies on. The O's indicate the positions of the Geese, and the X's indicate particularly critical positions for the Fox. When the Fox is in one of these places, we shall say that the Geese are *in danger*.

If the Geese follow our advice, they will play the game as follows. Most of the time they should play with their eyes closed, so that they will not be alarmed unnecessarily by the Fox's manoeuvres. We can offer them a guarantee that whenever they open their eyes the position will be like one of A, B, C, D, E, possibly left-right reflected, and that all they need do before closing their eyes again is see whether or not they are in danger. If not, they should make the moves indicated by the digits before the solidus (/) in Fig. 3, and if in danger, those indicated by the digits after the solidus. We also show by letters before and after the solidus which type

A, B, C, D, E of position will be seen next. The indicated moves can be made with eyes closed, since we can also guarantee that the Fox will never be in the way, and so the Geese need only open eyes again when the sequence has been completed and the position is once again one of A, B, C, D, E.

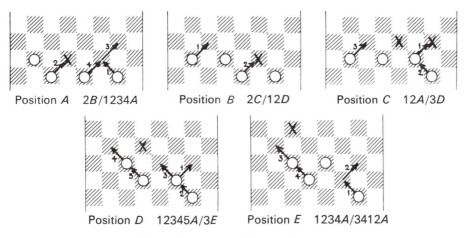

Position A 2B/1234A   Position B 2C/12D   Position C 12A/3D

Position D 12345A/3E   Position E 1234A/3412A

Figure 3. The Most Concise Strategy.

It is very easy to prove that the strategy works, when once the Geese have got into position A. As an example, we consider the position D. If the Geese have been behaving as we suggest, they can only have arrived at a position D from a position B or C in which they were in danger, and so the Fox can only be in one of a limited number of places. In fact he will be in one of X, Y, Z, T of Fig. 4.

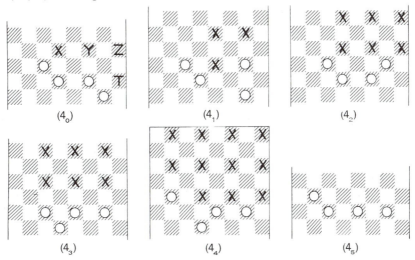

(4_0) (4_1) (4_2)

(4_3) (4_4) (4_5)

Figure 4. Analysis of Position D.

Now if the Geese are in danger (i.e., the Fox is at X), they make a single move and arrive immediately at position E. So we can suppose that the Fox is at Y, Z, or T, and the Geese are told to make moves 1, 2, 3, 4, 5. Figures $4_1, 4_2, 4_3, 4_4, 4_5$ show the position after each of these, and show why the next move in the sequence is legal, for in Figs 4_1 to 4_4 we have marked X in every possible place the Fox can be in just before the next move takes place. Since the position shown in Fig. 4_5 is of type A, this verifies that the strategy works from positions of shape D. Note that if the Fox were at T in Fig. 4_0, then he lost instantly after move 1, being trapped at the edge of the board.

We leave to the reader the corresponding discussion for positions of shapes A, B, C, E, noting merely that since E can be reached only from a position D with the Geese in danger, we need consider just two places for the Fox in position E, namely that marked X in Fig. 3, and the place two squares to the right of it. But for positions A, B, C the Fox may be on any square of the right color that is above the line of Geese.

How do the Geese start the strategy? The answer is that they can move into a position A on their first move, unless the Fox chooses the starting position F of Fig. 5. In this case, we advise them to make moves 1, 2, 3 and then open their eyes again to behold another anomalous position, G, from which we can put them back on track by giving them another sequence of three moves to be performed with eyes closed.

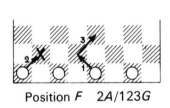   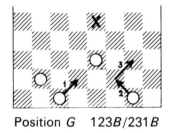

Position F 2A/123G   Position G 123B/231B

Figure 5. The Anomalous Starting Position.

Some Properties of Our Strategy

From the standard starting position, and indeed from any position except the anomalous position F of Fig. 5, our strategy leads only to positions in which the Geese never occupy places in either the leftmost or the rightmost column. This is interesting because most reasonably competent players like to move the Geese so that they straddle a horizontal row whenever possible. But if the Geese do this, a cunning Fox can force them into a wider variety of positions than occurs in our strategy, so that the Geese then need to know a lot more about the game to be certain of winning. In fact a competent Fox can force the Geese into positions of all the types A, B, C, D, E, no matter what winning strategy they adopt. However, only if they adopt our strategy is he unable to force them into any other position in which they need open their eyes because their action cannot be automatic. Since our strategy is the only one with as few as five positions in which the Geese need to take a decision, it is the unique *minimal* winning strategy.

Many people who think they know winning strategies can be caught out every now and then by a clever Fox who seduces them into an unfamiliar part of the game. In fact it takes considerable skill to play the Fox against ordinary players so as to exploit to the full any deficiencies in their knowledge. We can give few hints here beyond remarking that the Fox should stay near to the Geese and try to bring them into the middle of the board around him before stepping sideways to slip through any gap they may leave at either side. The best starting position is from the square near to the Geese and directly below the X of Fig. 2 (and so two squares right of X in position F), but position F itself is also useful.

What Is the Value of Fox-and-Geese?

So far, this chapter has been copied verbatim from the original 1982 edition of Winning Ways. That chapter went on to give a proof, which we described as "admittedly rather fluid," of the assertion that

"The value of Fox-and-Geese is $1 + 1/\mathbf{on}$."

We remained steadfast in that belief until we heard objections from John Tromp. We then also received correspondence from Jonathan Welton, who seemed to prove to somewhat higher standards of rigor that

"The value of Fox-and-Geese is $2 + 1/\mathbf{on}$."

Who was right? As often happens when good folk disagree, the answer is "both!" because it turns out that the parties are thinking of different things. The *Winning Ways* argument supposed an indefinitely long board, while Welton more reasonably considered the standard 8×8 checkerboard.

But why are we so interested in values?

Fox-Flocks-Fox

You can play FFF on an ordinary checkerboard, with the checker kings for the Foxes and four ordinary checker pieces of each color for the geese. The white geese move upward, the black ones downward, and the starting position is shown in Figure 6.

With best play, does this game continue forever, or is it a second-player-win in finitely many moves? What happens if the two Foxes are moved to an alternative symmetric pair of positions? [Answers are in the Extras.]

Towards Greater Precision

We unequivocally apologize for the fact that the statement of our "theorem" was just as fluid as its "proof"! In the rest of this chapter we shall try to make amends by giving precise statements and better proofs of both of these results, and of some new ones. Since the subtlest arguments are those for the Fox, we have found it convenient to take his viewpoint, and so have turned the board upside-down, putting the "Goosy" end at the top and the "Foxy" one at the bottom.

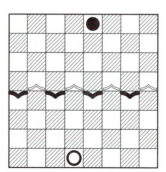

Figure 6. Fox-Flocks-Fox.

The Indefinite Board

How long should the board be? As regards the goosy end, it doesn't really matter: to win on either board in Figs. 7a and 7b, the Fox must get at least to the topmost row containing a Goose; but if he **is** able to do this then he **can** certainly win, by moving back and forth between this row and the one above, and so he needs no rows higher than these two (and of course, the Geese cannot use any such rows).

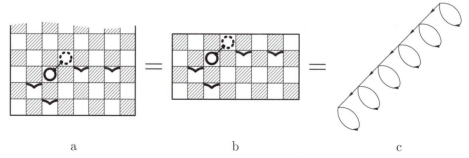

Figure 7. (a) Indefinite; (b) Definite; (c) 5 **off**.

What about the Foxy end? The length of the board at this end **does** affect the value, because many of his escapades need some backfield in which the Fox can maneuver. However, to choose any particular length merely complicates the play, so in the first instance we prefer to simplify the game by making the board indefinitely long at the Foxy end too, but to make the values precise by declaring, nevertheless, that our statement

"When the Fox gets to the level of the highest Goose, the value is **off**."

remains true for the indefinite board, too.

The next few sections are devoted to a proof that many values on the indefinite board are indeed exactly $1 + 1/$**on**, which we abbreviate to $1 + $**over**, or just 1**over**, following our usual conventions for omitting the $+$ sign. In tables and formulas, we shall often denote **over** by the symbol ε. Later we shall return to the distinctly more interesting results for definite boards.

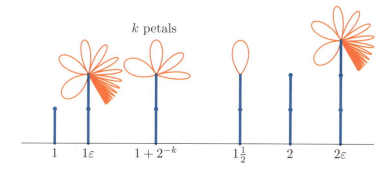

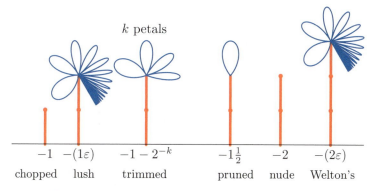

Figure 8. A Garden of Geraniums and Delphiniums.

How the Geese Survive Triumphantly

We show first that the Geese can survive forever in

$$\text{Fox-and-Geese} + \text{Lush Delphinium,}$$

where the latter is a Hackenbush flower consisting of a red stalk of length 2 supporting an infinity of blue petals. A picture of it can be found in the garden of Fig 8.

Then GOOSETAC, the Geese's Tactical Table shown in Fig 9, shows them what to do almost all the time. To follow it, remember that it shows only moves on the Fox-and-Geese board; every now and then players will *hack* at the delphinium flower instead, and so appear to pass on the Fox-and-Geese board. Obviously, if the Fox could pass freely but the Geese not, then he would be able to win just by waiting for them to be forced to move past him - but he can't, and he won't in fact be able to win.

The typical board in GOOSETAC is labeled by a letter and a number, the number being the total height of the Geese (Geese in the lowest rank having height 0), counted modulo 8.

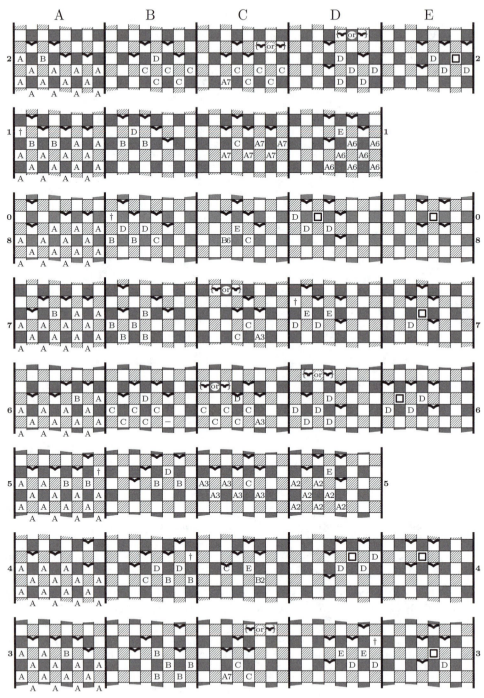

Figure 9. GOOSETAC

The four symbols ⌣ on each board represent the Geese; other symbols indicate possible locations for the Fox. A capital letter advises the Geese to move to the formation in the next row whose column is headed by that letter, unless it is followed by a digit n, when the next formation is in row n.

In the first column, the letters A continue indefinitely downwards. The Fox will avoid positions marked † because he can be trapped next move. It is easy to check that the lettered move is always possible, and leads to another formation of GOOSETAC in which the Fox is on a marked square, even if he hacks. If the Fox is on a boxed square at the Geese's turn, we would recommend them to resign if they have no more hack moves. Fortunately, when playing on the indefinite board plus the lush delphinium, we can assure them that they will never need to resign if they follow "GOOSESTRAT," their Strategic diagram in Fig. 10 on page 678.

The proof makes use of the fact that there are two types of position, *dark* ones, in which the Fox is in a row containing black squares, and *light* ones, when he is in a row containing gray ones. Any move by either player on the Fox-and-Geese board changes from one of these types to the other, and so is either a *darkening* move, changing from light to dark, or a *lightening* one, changing from dark to light.

Since the boxed squares are in the D and E columns, it is only the outward moves which pose potential difficulties for the geese. So GOOSESTRAT shows all non-inward moves for the Geese, and indicates for which of them we recommend the Geese to hack (if they still can). Of course each Fox move occurs at just one node.

In the absence of hack moves, all Geese moves would be of the same type; all darkening or all lightening. But the position could only get into column E by changing type twice, which would involve two hacks, necessarily by the Fox. So for positions in this column it is his turn and he is out of hacks, so he must have to move from the boxed square, and the Geese survive!

If the Geese were not allowed to hack, then the Fox could get to D0 having used only one hack (at B1), after which he could pass at D0 and they'd be stuck. But the lush delphinium, of value $-(\textbf{1over})$, provides the geese with a bountiful supply of hacks and so the geese can also survive at A2 and A6.

These arguments verify that the Geese can survive indefinitely in Fox-and-Geese plus the lush Delphinium, and so prove the inequality

$$\text{F\&G} \geq \textbf{1over},$$

where for definiteness the geese formation here is A2, and the Fox can be anywhere beneath the geese. To establish the opposite inequality, we'll soon turn our attention to the Fox. But we must first investigate

Eight Exciting Escapades

In Escapades 1–5, Figures 11 and 12, the initial locations of the four Geese are indicated by the usual symbols, and that of the Fox by a circle. The moves sensible players will probably make thereafter are shown by numbers, odd ones for the Fox and even for the Geese. The reader should play these moves out on the board, and verify that indeed they lead to escapes for the Fox.

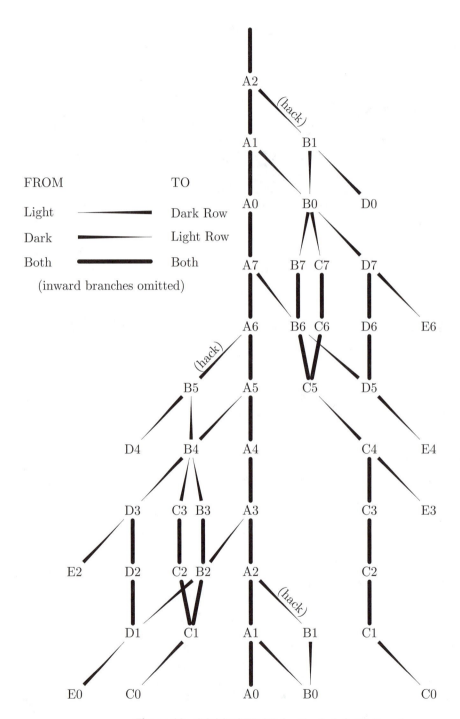

Figure 10. GOOSESTRAT for the Indefinite Board.

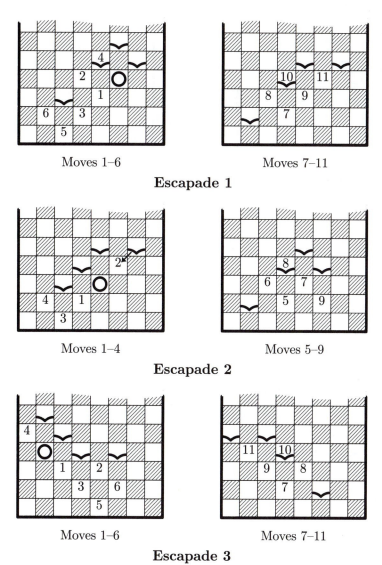

Moves 1–6   Moves 7–11

Escapade 1

Moves 1–4   Moves 5–9

Escapade 2

Moves 1–6   Moves 7–11

Escapade 3

Figure 11. Escapades 1–3.

Can the Geese find some other moves to prevent this? The reader who has played through these sequences will see that most of the Goose moves are forced to forestall the Fox from breaking through immediately. For instance if the Geese didn't make the move to the square marked 2 in Escapade 1, the Fox would win next move by taking this square himself. There are some cases where alternative Goose moves seem to offer some hope, but the careful reader will easily show that in fact none of them can prevent the Fox's escape.

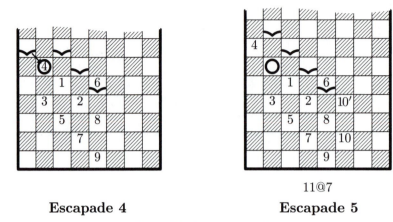

11@7

Escapade 4   **Escapade 5**

Figure 12. Escapades 4–5.

Fox moves: 1, 3, 5, 7, 9, b, d, f, h, j, ...
Goose moves: 2, 4, 6, 8, a, c, e, g, i, k, ...

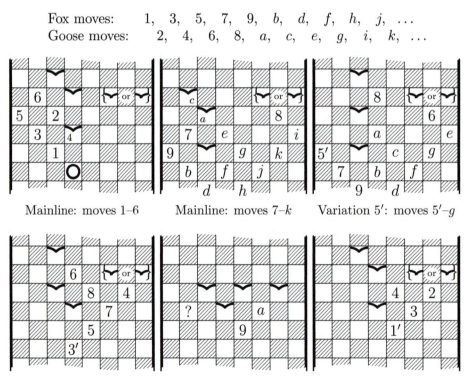

Mainline: moves 1–6   Mainline: moves 7–k   Variation 5′: moves 5′–g

Variation 3′: moves 3′–8   Variation 3′: moves 9–a   Variation 1′: moves 1′–4

Figure 13. Escapade 6.

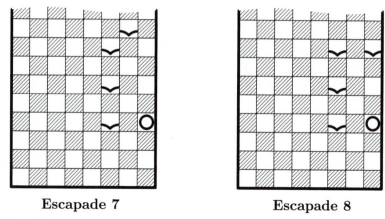

Escapade 7   **Escapade 8**

Figure 14. Escapades 7–8. In each Case, Geese to Move. Can Fox Escape?

These escapades work not only on the indefinite board, but on definite ones provided the lower boundary gives Fox enough room in his backfield. The ones we have drawn show the smallest backfields that will suffice.

Escapade 6, Fig. 13, is a more challenging position in which clever geese are able to foil any attempt the Fox makes to escape. As some of these lines of play last for more than 10 moves, we number moves after 9 with the letters $a, b, c...$ The reader who plays through each of these variations may discover a key stratagem which the geese can also apply in some other positions. The southernmost goose remains in the same location for many moves, forcing the Fox either to retreat, or to commit himself to trying to pass on either the east side or the west side. The northern goose begins on a location from which he is able to move to help reinforce either the east or the west. But this goose must not be committed prematurely. In several variations, the northern goose arrives on the eastern or western scene just in time to block the Fox's escape. In variation $3'$, the geese might choose to answer move 9 by playing at "?" instead of at "a." Although this gives a solid formation from which the Fox cannot escape, it is often not as good a move as "a", for subtle reasons that will become apparent in a later section.

What if the Fox moves first from starting position of Escapade 6? We'll give the answer in the Extras. Escapades 7 and 8 in Fig. 14 offer more exercises for the experienced player; their solutions appear in the Extras.

Escapades 1–5, and some easier ones we'll take for granted, will now help us prove that it's also true that the Fox can survive indefinitely in Fox-and-Geese plus the lush Delphinium, so that F&G = 1**over**. They are an essential part of our tactics and strategy for the Fox.

Tactics and Startegy for the Fox

FOXTAC, the Fox's Tactical Table, is somewhat like GOOSETAC. The Fox is always on a circled position in each board. If this has a ⊙ inside, it is a bullseye from which he can escape

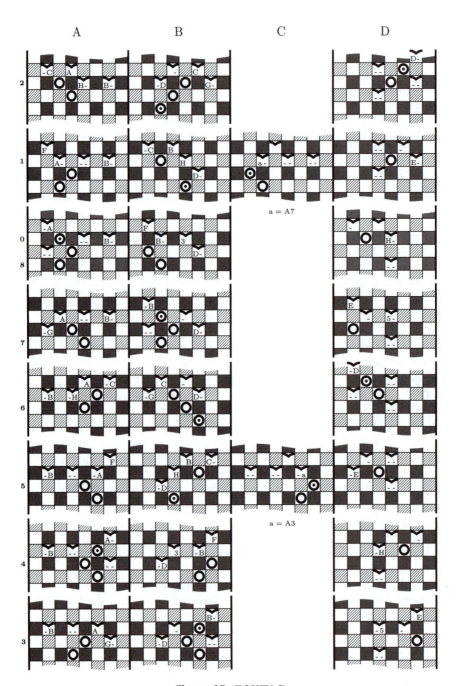

Figure 15. FOXTAC.

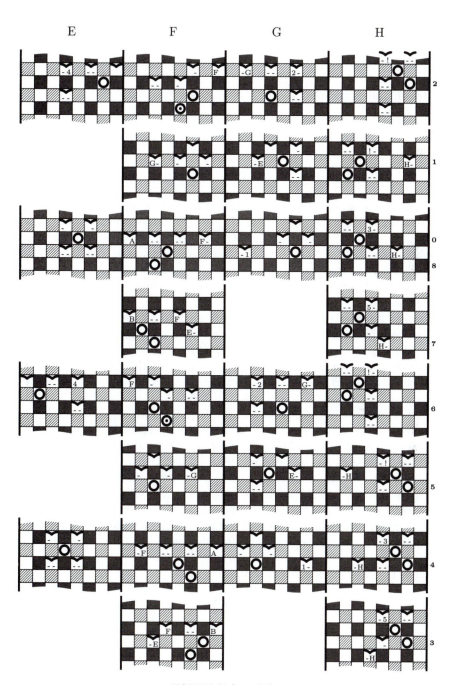

FOXTAC (cont'd).

even if it's the Geese's turn, by an escapade so easy that we don't need to give any further details. For each diagram, if it's the Fox's turn, he should, in order:

1. escape, if on a bullseye, or

2. move to a bullseye, and then escape. Otherwise

3. move to a circle, else

4. hack, and otherwise

5. resign.

We shall show that when playing with the lush Delphinium, the last possibility won't happen. To see this, one must check that all possible Geese moves (indicated by symbols at their feet) lead back into FOXTAC, except for those indicated by −, from which the Fox escapes easily, and those indicated by single digits or exclamation points, from which the Fox escapes excitingly. Digits correspond directly to escapades (e.g., the digit "2" refers to Escapade 2; exclamation points indicate other exciting escapades which are very similar to Escapades 3, 4 and 5.

Notice that no letters except H appear in the H column of FOXTAC. So if the Fox stays on the circled locations in any H formation, he can escape if they move to any formation outside this column. If the geese hack, the Fox can respond with another hack. If the geese refuse to resign, they must eventually move to a formation from which the Fox can escape. The reader can verify that the FOXTAC table is closed except for those very bad Goosemoves to Column H.

When it is Goose's turn, if the Fox hasn't already won, he stays on a circled location on one of the diagrams in FOXTAC. One purpose of FOXSTRAT (Fig. 16 on page 685) is to show that when playing with a lush delphinium, (and having started at an appropriate place), the Fox will not find himself at an *isolated* circle with no hack move available. Pairs of adjacent circles give him no trouble, since he can move between them.

Now the positions having isolated circles are in the outer pair of columns in FOXSTRAT, and are the only ones at which the Fox will hack, so the first time he is in such a fix, both his hack moves will be available, and he can use his first hack. Also, in these positions, there is no Goose move to a non-circled node, so after this hack, the Geese must move to another such position (or out to H). There are two cases for their moves, which will be vertical in the figure, say from Y to Z or from X to Y to Z:

The Fox's two hack moves clearly enable him to survive in the first case. In the second case, they also suffice, because both moves are of the same type (darkening or lightening), showing

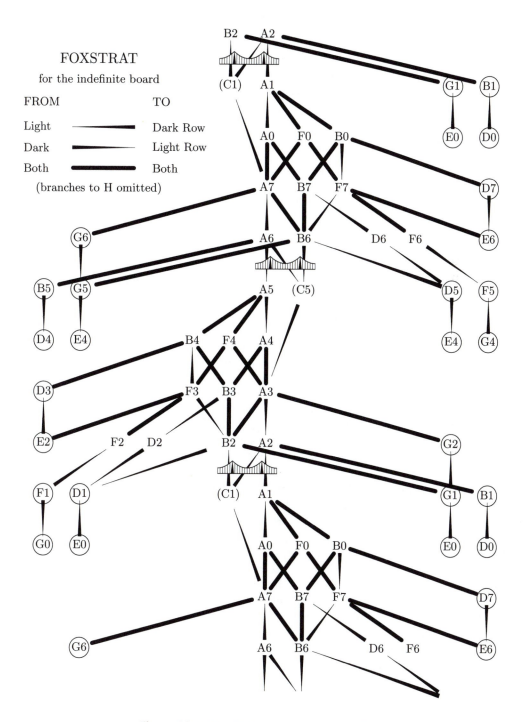

Figure 16. FOXSTRAT for the Indefinite Board.

that he won't be at the middle node Y at his turn, and can therefore hack again at Z. After this second hack, the Geese have run out of moves.

This, together with the many easy escapade checks we've omitted, shows that the Fox can also survive indefinitely when playing Fox-and-Geese on the indefinite board along with our flower, and so establish the inequality

$$\text{F\&G} \leq \textbf{1over}.$$

So, combining the two opposite inequalities, we have the exact evaluation

$$\text{F\&G} = \textbf{1over} = 1 + 1/\textbf{on},$$

where, to be precise, we remind the reader that "F&G" refers to any position in which the geese are in formation A2 or A6, and the Fox is located anywhere below the geese.

Fox's Play on Definite Boards

On definite boards, the rows are numbered upwards from 0, in octal, and each diagram shown in GEESETAC, GEESESTRAT, and FOXSTRAT is now labeled with a letter followed by an octal number of at least two digits. (To avoid ambiguity, we insert a leading zero before each height less than 8.) Then A2 represents any of the formations A02, A12, A22, A32,... At sufficiently large heights, the graphs shown in GOOSETAC, GOOSESTRAT, FOXTAC, and FOXSTRAT all remain locally valid, with the rows formerly cyclically labeled mod 8 being replaced with a long single sequence continuing on downward. These figures all remain accurate if the height is sufficiently large to ensure that Fox's escapes still work. When such an F&G position is played in conjunction with a trimmed Delphinium, the geese seek to push the Fox down towards the bottom of the board without hacking any more than necessary, while the Fox seeks to compel them to hack as often as possible. GOOSESTRAT showed how the Geese can push downwards without hacking anywhere except at most once at each formation of type A2 or A6. We will now show how FOXSTRAT implies that he can compel the Geese to hack at least this often. The bridges shown in FOXSTRAT are the key to proving this.

The Geese can move under the bridge that separates A1 and C1 from A2 and B2 only by playing a darkening move. After several more moves the Geese must either play to a circled position (from which Fox can win as on the indefinite board) or arrive at A6 or B6. Since our Fox does not hack from any of the positions in FOXSTRAT which are not circled, unless the Geese have hacked again very recently, they are now faced with the need to move from a light position. Since all moves under the bridge from A6 and B6 are lightening, the geese are unable to pass under this bridge without paying another hacking toll. Continuing this argument, whenever the geese cross under one bridge in FOXSTRAT, they will be forced to pay at least one hacking toll before they are able to cross under the next. And since the accompanying trimmed Delphinium has a stem of length two, the Fox can defeat any attempt by the Geese to evade these tolls by playing to any of the nodes which are circled in FOXSTRAT.

So Fox can continue this strategy as long as the position is high enough for his escapade threats to work. But since the height decreases, this assumption eventually must fail. The first failure occurs in H23. If the geese move from H27 to Escapade 5, the Fox escapes as shown there, but if the geese play the same way on the reflexion at H23, the Fox fails to escape because of insufficient backfield. Thus in FOXTAC, the goose move from formation D24 to H23 becomes playable. The template shown in FOXSTRAT provides only an incomplete representation of the graph relevant to the island containing A31, A30, A27, A26, B25, and D24, so it is no longer clear that two red stem branches are enough for Fox to win after Geese cross under the bridge from A32, B32 to A31 and C31. However, above this bridge, there are no such problems. All escapades work just as in the indefinite case, and if the Geese play to any circled formation ABOVE A32, B32, then any Hackenbush flower with a red stem of length at least two is sufficient for the Fox to win. In particular, playing from the higher circled *critical* position of

$$A32, A36, A42, A46, ...,$$

Fox playing second can win against a flower of value

$$2, \quad \frac{3}{2}, \quad \frac{5}{4}, \quad \frac{9}{8}, \quad ...,$$

respectively, and so we have bounds on the values of all such positions. To show that these bounds are tight, one of the things we will need to verify is that the Geese moving second can also win from the higher critical position A32 $-$ 2, a task we defer to the Extras. However, once this is shown, GOOSESTRAT provides a simple inductive argument that Geese moving second can also win from each of the critical positions A36 $-\frac{3}{2}$, A42 $-\frac{5}{4}$, A46 $-\frac{9}{8}$, ... So, subject only to verification of the case of height 32 (octal) (when $k = 0$) we have proved that

$$FOX \& GEESE = 1 + 2^{-k},$$

whenever the board is in any of the critical A positions of height $4k + 32$ (octal). We verified this important case computationally using Siegel's *cgsuite* described in the Extras.

The Scrimmage Sequence

The formations in column A play a central role in both GOOSESTRAT and FOXSTRAT. Like the fundamental ground game advocated by the legendary football coach Woody Hayes, these formations spread the geese out horizontally, along a *line of scrimmage* parallel to their goal line at the bottom of the board. They push the Fox forward, one row per hack. So we call these solid formations A7, A6, A5, A3, A2, and A1 the *scrimmage formations*. Using Siegel's *cgsuite*, we computed the values of all positions with the geese in a scrimmage formation on all boards of sizes up to 16 × 8. Table 1 shows the values for those positions in which the Fox is in either location circled in FOXTAC . These values are tabulated according to a single parameter, n, called the *altitude*. This altitude is defined as the sum of the row-ranks of all five animals. It is most conveniently expressed in decimal.

For all values of n exceeding the lower bounds shown at the bottom of the table, this same sequence also gives the value of the Fox in either circled position of formations F0 and B0. All of these formations correspond to nodes near the centre of the GOOSESTRAT and FOXSTRAT figures. In prior sections, we evaluated *critical positions* which all have altitudes

units→ tens ↓	9	8	7	6	5	4	3	2	1	0
...										
50s	$\frac{33}{32}$	$\frac{33}{32}*$	$\frac{33}{32}$	$\frac{17}{16}*\mid\frac{33}{32}$	$\frac{17}{16}*$	$\frac{17}{16}$	$\frac{17}{16}*$	$\frac{17}{16}$	$\frac{9}{8}*\mid\frac{17}{16}$	$\frac{9}{8}*$
40s	$\frac{9}{8}$	$\frac{9}{8}*$	$\frac{9}{8}$	$\frac{5}{4}*\mid\frac{9}{8}$	$\frac{5}{4}*$	$\frac{5}{4}$	$\frac{5}{4}*$	$\frac{5}{4}$	$\frac{3}{2}*\mid\frac{5}{4}$	$\frac{3}{2}*$
30s	$\frac{3}{2}$	$\frac{3}{2}*$	$\frac{3}{2}$	$2*\mid\frac{3}{2}$	$2*$	2	$2*$	2	$2\varepsilon\mid2$	2ε
20s	2ε	2ε	2ε	2ε	2ε	2ε	2ε	2ε	2ε	2ε
10s	2ε	2ε	2ε	2ε	2	$2*$	2	NOTE	$\frac{5}{2}*$	$\frac{5}{2}$
00s	$3*\mid\frac{5}{2}$	$2*$	2	$3\mid2$	3	NONE	1	0	NONE	NONE

NOTE: Ambiguous; A13 $= \frac{5}{2}*\mid2$, but A12 $= \frac{5}{2}$.

Values of A3, A2, A1 as function of altitude in decimal
and of F0 (if altitude > 12) and of B0 (if altitude > 32).

Table 1. Values in the Scrimmage Sequence, $\mathbf{s}[n]$; $n =$ Altitude.

congruent to 2 mod 5, and we proved the correctness of that column of the scrimmage sequence. Omar may wish to extend these arguments to prove the correctness of the scrimmage sequence for all large values of n.

The scrimmage sequence is naturally partitioned into three regions: a **high** region, "**Welton's region**" of value 2**over**, and a **low** region. Each of these regions is discussed further in the Extras.

Values of the Initial Positions

On the $J \times 8$ board, the game conventionally begins from an initial formation in which all four geese occupy the highest row. Usually the Fox starts on the lowest row. In decimal, the height and initial altitude are both $4J - 4$. But in some variations of the game, the Fox is allowed to pick another starting location, sometimes subject to one or more constraints. When $8 < J < 17$, we noticed that the values of *all* positions on the bottom $J - 7$ rows of the board belong to the *early sequence* shown in Table 2.

Based on considerable computational evidence, we close this chapter with the assertion that, with appropriate modifications and refinements detailed in the Extras, this early sequence provides the *precise* values for *all* positions in the initial formation of a $J \times 8$ board for *every* $J > 7$. The Extras contain more details about our asserted values for all Fox positions with geese in their initial formation on the top row of any sized board.

Rather than the usual route of starting with small numbers and working upwards, we have instead come here via a true top-down route: we started at infinity and worked downwards. Part of the reason this has succeeded is that the asymptotic values include many numbers,

units→ tens ↓	9, 4	8, 3	7, 2	6, 1	5, 0
. . .					
lo 60s	$\frac{33}{32}*$	$\frac{33}{32}$	$\frac{33}{32}*$	$\frac{17}{16}\mid\frac{33}{32}*$	$\frac{17}{16}$
hi 50s	$\frac{17}{16}*$	$\frac{17}{16}$	$\frac{17}{16}*$	$\frac{9}{8}\mid\frac{17}{16}*$	$\frac{9}{8}$
lo 50s	$\frac{9}{8}*$	$\frac{9}{8}$	$\frac{9}{8}*$	$\frac{5}{4}\mid\frac{9}{8}*$	$\frac{5}{4}$
hi 40s	$\frac{5}{4}*$	$\frac{5}{4}$	$\frac{5}{4}*$	$\frac{3}{2}\mid\frac{5}{4}*$	$\frac{3}{2}$
lo 40s	$\frac{3}{2}*$	$\frac{3}{2}$	$\frac{3}{2}*$	$2\mid\frac{3}{2}*$	2
hi 30s	$2*$	2	$2*$	2ish	2ish
lo 30s	2ish	2ε	2ε	2ε	2ε
hi 20s	2ε	2ε	2ε	2ε	

Table 2. Values of *All* Early-Stage Positions, $e[n]$; n =Altitude.

and the temperatures of the other positions in the scrimmage sequence tend to increase with decreasing altitude. So, the route we have taken may still be viewed as "bottom-up" from a thermal perspective. We started by building a foundation consisting of important stable positions and then related other nearby positions to them.

Extras

Sophisticated New Software

All of the calculations in the Fox and Geese chapter of the first edition of Winning Ways were done by hand. Indeed, virtually all calculations in all volumes of that edition were done by hand, the only exceptions being some very specific nimber recursions such as those used to explore Grundy's game (WW Chapter 4) and certain sequences of Kaylesvines (WW Chapter 16).

David Wolfe introduced a much higher level of computer sophistication to the combinatorial game theory community in the early 1990s. His pioneering toolkit is able to compute and manipulate canonical forms symbolically. It understands the basic concepts of numbers, nimbers, infinitesimals, atomic weights, cooling, heating, overheating, and thermography. It is interactive and relatively easy to use. Wolfe's toolkit supports the analysis of several specific games, including Domineering and Clobber. It also provides a platform which others have extended to other specific games, including Toads & Frogs (by Jeff Erickson), and Konane (by Michael Ernst).

The source code has long been publicly available and accessible online on Wolfe's website. By 2002, it had attracted hundreds of users, all running on Unix-based machines.

In the fall of 2002, Aaron Siegel began an ambitious effort to write a new suite of game theory software from scratch. His programs are all written in Java, with a great degree of generality and portability. He has also given considerable attention to software engineering issues such as cache management, allowing his toolkit to handle larger problems faster. The resulting suite supports all of the features in Wolfe's toolkit and a number of important new ones, including the capability of handling loopy games like Fox and Geese. Siegel devised and developed his own new algorithms for handling such games, based on the concepts and theories found in WW Chapter 11.

We have used Siegel's program to crosscheck some of the results presented in this chapter, and to discover some of the others. The current [spring 2003] version of Siegel's toolkit can compute the canonical form of any small collection of Fox and Geese positions on a 14×8 board in a few minutes, and on a 16×8 board in about 90 minutes, running on a 2.4 GHz Pentium 4 with 512 MB of RAM. With our own efforts to analyze and understand Fox and Geese, we have become one of the first users of Siegel's new *cgsuite*. Documentation and diagnostics are being upgraded continually. David Wolfe has contributed advice, and a group in New Zealand led by Michael Albert is working on some additional graphical components.

Some readers will no doubt be eager to use this wonderful toolkit for further explorations of both Fox and Geese and of many other games. It is publicly available and can be accessed via http://cgsuite.sourceforge.net

One of our first uses of this toolkit was to confirm the fact that the value of the *critical position* with Fox in the higher circled location of A32 is 2, thus completing the proof of one of our new theorems.

690

	7 and 3				6 and 2				5 and 1		4 and 0			
	A	B	F	H	A	B	F	H	A	H	A	B	F	H
lo 40s	$\frac{5}{4}*$	$\frac{5}{4}\vert-$	$\frac{9}{8}$	$-$	$\frac{5}{4}$	$\frac{5}{4}$	$1\vert-$	$-$	$\frac{3}{2}*$	$-$	$\frac{3}{2}*\vert-$	$\frac{3}{2}$	$\frac{3}{2}$	$-$
	$\frac{5}{4}$	$\frac{5}{4}$	$2\vert\frac{5}{4}\Vert\frac{9}{8}$	$-$	$\frac{3}{2}*\vert\frac{5}{4}$	$\frac{5}{4}\vert-$	$-$	$-$	$\frac{3}{2}$	$-$	$\frac{3}{2}*$	$\frac{3}{2}*$	$\frac{3}{2}*$	$-$
hi 30s	$\frac{3}{2}*$	$\frac{3}{2}\vert-$	$\frac{5}{4}$	$-$	$\frac{3}{2}$	$\frac{3}{2}$	$1\vert-$	$-$	$2*$	$-$	$2*\vert-$	2	2	$-$
	$\frac{3}{2}$	$\frac{3}{2}$	$2\varepsilon\vert\frac{3}{2}\Vert\frac{5}{4}$	$-$	$2*\vert\frac{3}{2}$	$\frac{3}{2}\vert-$	$-$	$-$	2	$-$	$2*$	$2*$	$2*$	$-$
lo 30s	$2*$	$2\vert-$	$\frac{3}{2}$	$-$	2	2	$1\vert-$	$-$	2ε	$-$	$2\varepsilon\vert-$	2	2ε	$-$
	2	2	$2\frac{1}{2}\vert2\Vert\frac{3}{2}$	$-$	$2\varepsilon\vert2$	$2\vert-$	$-$	$-$	2ε	$-$	2ε	$2*$	2ε	$-$
hi 20s	2ε	$2\vert-$	$\frac{3}{2}$	$-$	2ε	2	$1\vert-$	$-$	2ε	1	$2\varepsilon\vert-$	2	2ε	$1*$
	2ε	2	$2\frac{1}{2}\vert2\Vert\frac{3}{2}$	$-$	2ε	$2\vert-$	$-$	$-$	2ε	$1*$	2ε	$2*$	2ε	1
lo 20s	2ε	$2\vert-$	$\frac{3}{2}$	1	2ε	2	$1\vert-$	0	2ε	1	$2\varepsilon\vert-$	2	2ε	$1*$
	2ε	2	$2\frac{1}{2}\vert2\Vert\frac{3}{2}$	$2\varepsilon\vert1$	2ε	$2\vert-$	$-$	$3\vert2\varepsilon\Vert0$	2ε	$1*$	2ε	$2*$	2ε	1
hi 10s	2ε	$2\vert-$	$3\vert2$	1	2ε	2	$3\vert-$	0	2	2ε	$2\vert-$	2ε	$2*$	2
	2ε	2	2	$2\varepsilon\vert1$	2ε	$2\vert-$	$-$	$2\varepsilon\vert0$	$2*$	2ε	2	$2\varepsilon\vert2*$	2	$3\vert2\varepsilon\Vert2$
lo 10s	2	$\frac{3}{2}\vert-$	$2\vert\frac{3}{2}$	1	$2\frac{1}{2}$	$\frac{3}{2}$	$2\vert1\varepsilon$	0	$2\frac{1}{2}$	2	$2*\vert-$	$2\Uparrow$	2	1
	$2\frac{1}{2}*\vert2$	$2*\Vert\frac{3}{2}\vert-$	$\frac{3}{2}$	$2\varepsilon\vert1$	$2\frac{1}{2}*$	$\frac{3}{2}\vert-$	$5\vert1\varepsilon\Vert-$	$1\varepsilon\vert0$	$3*\vert2\frac{1}{2}$	$3\vert2$	$2*$	$2\uparrow*$	$2\varepsilon\vert1$	$2\vert1$
hi 00s	$2*$	$2\vert-$	$\frac{3}{2}$	0	2	2	1ε		3	1	$1\vert-$	1ε	2	0
	2	$2\Vert2\vert-$	$3\vert2\Vert\frac{3}{2}$	$1\vert0$	$3\vert2$	$3\vert2$				$1\varepsilon\vert1$				$\varepsilon\vert0$
lo 00s	1	$0\vert-$	1											

Table 3. Values of Fox's Tactical Diagrams with Double Circle. (Value at Higher Circle Listed above Value at Lower Circle.)

	7 and 3	6 and 2			5 and 1					4 and 0		
	D	D	E	G	B	C	D	F	G	D	E	G
lo 40s	1	$1\vert-$	0	0	1	$\frac{3}{2}*\vert-$	1	1	1	0	0	0
hi 30s	1	$1\vert-$	0	0	1	$2*\vert-$	1	1	1	0	0	0
lo 30s	1	$1\vert-$	0	0	1	$2\varepsilon\vert-$	1	1	1	0	0	0
hi 20s	1	$1\vert-$	0	0	2	$2\varepsilon\vert-$	1	1	1	$1\uparrow$	0	0
lo 20s	1	$1\vert-$	$2\vert1$	0	2	$2\varepsilon\vert-$	1	1	1	$\frac{5}{4}$	0	0
hi 10s	$1\frac{1}{2}*$	$1\vert-$	2ish	0	$2*$	$2\vert-$	1	3	1	2	0	2
lo 10s	2ε	$1\vert-$	1ε	$2*\vert\frac{3}{2}$	2	$2*\vert-$	1	2	1	1	0	1
hi 00s	$1\Uparrow$	$1\vert-$	$\frac{1}{2}$	1	1		1	1	1		0	0

Table 4. Values of Fox's Tactical Diagrams with Single Circle.

Values of FOXTAC Positions

Tables 3 and 4 show the values of most circled positions in FOXTAC, as computed by Siegel's *cgsuite*. In most of these formations, these circled positions are among the best locations for a Fox who has neither escaped nor is threatening an imminent escape. All of the exceptions, shown in the two boldfaced values in each of the columns of Table 5 occur only on rather short boards.

altitude mod 5	3		2	0
	D	F	E	B
...		....		
hi 20s	1	$2*$	$1*$	2
lo 20s	1	$2*$	$1*$	2
hi 10s	$2\varepsilon\|\frac{3}{2}$	$\mathbf{2*}$	$\mathbf{1*}$	$\mathbf{1*}$
lo 10s	2	$\frac{3}{2}*$	$\frac{3}{2}$	**2**
hi 00s	1	$2*$	1ε	2

Table 5. Values at Locations toward Centre from Higher Circle. (Each Boldfaced Entry $<$ Circled Outsider)

These tables prove very helpful in the next section.

Regions of the Scrimage Sequence

High Region

We have observed contests between two strong players starting from one of the very high scrimmage positions played in conjunction with the negative of a Hackenbush string of equal or nearly equal value. After the first five pairs of consecutive moves, the position reverts to a reflected translation of its initial position. The five geese moves consist of four advances and one pass; the five Fox moves consist of two advances and three retreats. The formation's height has decreased by four; the position's altitude has decreased by five, and the geese have used up one of their Hackenbush petals. Why can good players do no better than this? Because for each relevant formation of the Geese, the locations circled in FOXTAC are as minimal as any others. So the Fox loses nothing by staying within the circled positions. And if he does so, our prior FOXSTRAT arguments have shown that the geese cannot play to any formation not shown in FOXSTRAT without allowing the Fox to escape.

The stops of the values which appear within the high region are all of the form

$$1 + 2^{-k}.$$

The transition between the high region of the scrimmage sequence and Welton's region of value **2over** happens at altitude 31. This position occurs with the geese in formation A32 (octal) and the Fox in the lower circled position. We view the value of this position as a switch of temperature **over**. When played in conjunction with the Hackenbush string of value -2, the mover can win. If Fox moves first, he advances and the geese are then unable to pay the toll necessary to go under the bridge, so they must move outward from the central nodes to a position from which they will lose within at most another few moves. But if the geese are able to play from this position, they go under the bridge to enter the next region.

Welton's Region of Value 2**over**

This region was explored extensively by Welton. All values in this region exceed the number 2 by the loopy positive infinitesimal ε called "over." This infinitesimal exceeds any finite integer multiple of UPs. Its persistence for so many altitudes may be attributed, at least partly, to the fact that it satisfies the following broad range of equalities:

$$\varepsilon = 0|\varepsilon = \varepsilon|\varepsilon = \varepsilon + \varepsilon = \varepsilon*$$

Evidently, ε is an idempotent. Although it would be absorbed by any of the idempotent thick stacks of coupons discussed by Berlekamp [2002], it has sufficient vim to absorb any conventional infinitesimal such as ⇑ or ⤊.

In Welton's region, the appropriate Hackenbush companion is Welton's Delphinium, which has an infinite number of blue petals atop a red stem of length 3. Our strategic graphs for positions in this region differ only very slightly from the templates shown in FOXSTRAT and GEESESTRAT, but the difference is quite important: Column H is no longer disastrous for the Geese. This is because the bottom of the board is now close enough that the Fox can no longer escape by Escapade 5. So the line of play from formations A26 to B25 to D24 to H23 is not disastrous for the geese, but only mildly disadvantageous. Fig. 17 shows the portion of FOXSTRAT between the bridge entering C31, A31 and the bridge exiting from A26. The first major difference from the template of FOXSTRAT is the presence of some new nodes H25, H24, etc.

Another major difference is that the appropriate Hackenbush companion, Welton's Delphinium, has a longer stem. So there are many more nodes from which the Fox can win by hacking. All such nodes have been circled. In both Welton's region and the low region, we advise the Fox to consult the value tables at every move, and if he can win by hacking, he should prefer to do that rather than to play. Some examples of positions in which he MUST be sure to do this are shown in Fig 18.

By using Tables 3, 4, and 5, we can easily construct the correct graph corresponding to FOXSTRAT. The region of FOXSTRAT between the bridge into A25 and the bridge exiting A22 is an exact translated reflexion of Fig. 17. The region between the bridge into A21 and the bridge leaving A16 is almost an exact translation, but it has exactly one exception: the

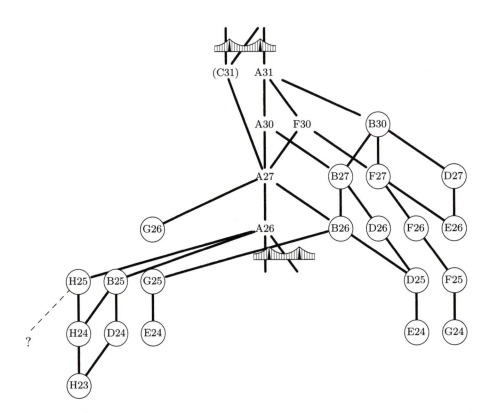

Figure 17. Fox's Strategic Landing Plan (Darkening and Lightening Omitted) .

node H15 is not circled. The next lower region, beginning with A15 is quite different and lies entirely outside of the **2over** region.

We gain the most insight by watching a contest between two expert players starting with Welton's Delphinium added to an F&G position in which the Geese are in the scrimmage position A31. Fox plays according to a slightly amended version of the FOXSTRAT we have studied previously; the only difference being that before playing, he should consult the tabu-

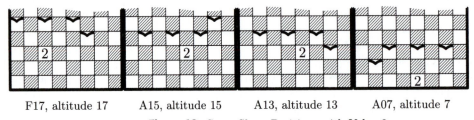

F17, altitude 17   A15, altitude 15   A13, altitude 13   A07, altitude 7

Figure 18. Some Short Positions with Value 2.

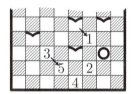

Figure 19. Journey from H15 to Fox's Safe Dancing Haven.

lated values and HACK instead of playing whenever that would permit him to win, as it now often does because the three-branch stem of the Hackenbush flower provides him with more possible hacks. Smart geese refuse to move into any off-centre formations. They instead elect to play either a passive or an aggressive strategy. If passive, they simply remain in a formation such as A31, A27, or A26, and hack at every turn. Fox can do no better than engage in his two-step dance, back and forth between his pair of circles. The total position repeats again and again, and the game is drawn.

More aggressive smart geese may attempt to force the Fox down to positions of lower and lower altitudes. They can succeed in pushing from A31 thru A30 to A27 to A26. Although they may then need to pay a toll to get under the bridge to A25, this poses no problem because Welton's Delphinium has an infinite number of blue petals. After playing one of them as a toll, the geese push on under the bridge to A25 and then continue thru A24, A23, and A22, then paying another toll before moving on to A21. However, after crossing under this bridge, they find themselves in new territory. Several more of Fox's former escapades now have insufficient backfield, and in particular, formation H15 is better for the geese than the *central* formation A15. So the dominant strategy for the aggressive geese is to push from formation A21 thru A20, A17, and A16, to H15 and then, after hacking if necessary, off the charts thru Fig. 19 to Fig. 20. But this is as far as the smart aggressive geese can go; they are, at last, compelled to resort to the passive behavior of hacking at every turn while the Fox remains alive and active in his final safe dancing haven. If the geese play any more aggressive moves from that position + Welton's Delphinium, Fox can win. So we may regard *all* of the positions in Welton's region as predecessors of this simpler position we call Fox's Safe Dancing Haven.

In almost all of the positions in Welton's region, the geese have a dominant move to another position of value 2**over**. However, canonical play requires them only to reach a value >= 2. Figures 21 and 22 show how a competent human player can do this with only a very few minor amendments to GOOSESTRAT and GOOSETAC.

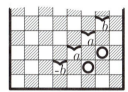

Value $= 2\varepsilon$; $a = 2$; $b = 1\varepsilon$.

Figure 20. Fox's Safe Dancing Haven.

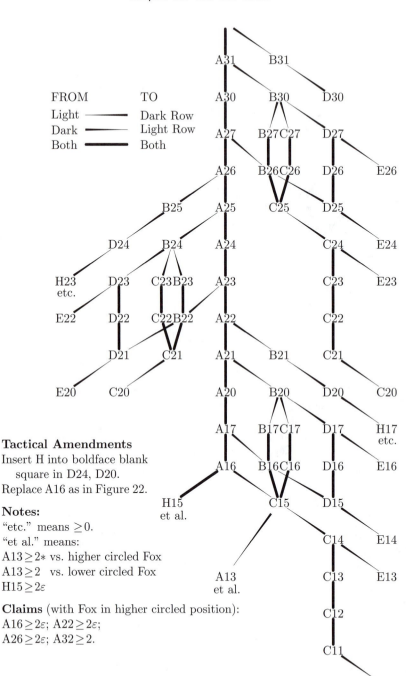

FROM   **TO**

Light ——— Dark Row

Dark ——— Light Row

Both ━━━ Both

Tactical Amendments

Insert H into boldface blank
  square in D24, D20.
Replace A16 as in Figure 22.

Notes:

"etc." means ≥ 0.

"et al." means:

A13 $\geq 2*$ vs. higher circled Fox

A13 ≥ 2 vs. lower circled Fox

H15 $\geq 2\varepsilon$

Claims (with Fox in higher circled position):

A16 $\geq 2\varepsilon$; A22 $\geq 2\varepsilon$;

A26 $\geq 2\varepsilon$; A32 ≥ 2.

Figure 21. Geese's Strategic Landing Plan (Inward Branches above C13 Omitted).

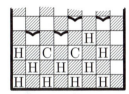

Figure 22. Tactical Amendment of Geese's Landing Diagram A16 .

Lower "Delta" Region

Like a major river that fans out into a delta before flowing into the sea, good lines of play fan out into many variations in the lower region. Not only do the scrimmage values in Table 1 on page 688 no longer apply to formations of types F0 and B0, but there is even one case (at altitude 12) where A12 and A13 have different values. As shown in Table 5 on page 692, there are also now some positions with values smaller than the locations circled in the templates shown in FOXTAC.

In short, patterns that prevailed at higher altitudes fan out into a plethora of special cases in the lower region. Fortunately, this region is not only small enough that all values can be computed very quickly by Siegel's *cgsuite*, it is also small enough to have been mastered by some human experts without any computer assistance.

Initial Values

Even when the geese are arranged in one of their solid scrimmage formations, if they are near enough to the bottom of the board, it is no longer clear which are the best locations for the Fox. So it may be useful to tabulate values of *all* Fox locations, as shown in Figs. 23 and 24 with the geese in their initial F4/F0 formation at the top of a board. Notice that these values assume an indefinite top of the board, as in Fig. 7a, so they provide answers to questions such as the Fox-Flocks-Fox problem of Fig. 6. Allowing the Fox to be trapped on the *top* row of a fixed-sized $J \times 8$ board would make some small-board initial positions more advantageous to the Geese.

Some interesting patterns will appear on larger boards which aren't yet apparent in Welton's region.

Parity in the High Scrimmage Region

It is useful to regard all of the nodes in the FOXSTRAT and GEESESTRAT graphs which have values specified by the scrimmage sequence as forming the *high central region*. Nodes circled in FOXSTRAT, from which a competent Fox will not allow the geese to return to the central region, are viewed as having left this region.

The stops of the values which remain within the high scrimmage region are all of the form

$$1 + 2^{-k}.$$

5×8 BOARD

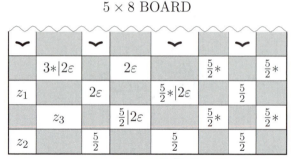

where $z_1 = 5* \,\|\|\| \, 4, 4* \,\|\| \, \frac{7}{2}|3\|2\varepsilon \,\|\|\| \, 3*|2\varepsilon$,
$z_2 = 4, 4* \,\|\| \, \frac{7}{2}|3\|2\varepsilon$, and $z_3 = \frac{7}{2}|3\|2\varepsilon$

4×8 BOARD

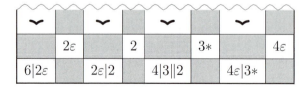

where $z_1 = 6\varepsilon \| 3|2$

3×8 BOARD

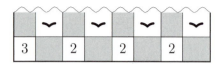

2×8 BOARD

| 3 | | 2 | | 2 | | 2 | |

Figure 23. Initial Values in the Low Region.

It is useful to regard this number has having the same PARITY as the integer k. Although the value of 1 is a canonical Left follower of these stops, its appears only on outer nodes external to the high central region. So within the high central region, every position has an even or odd value corresponding to the parity of its height. Even positions have only odd followers, and odd positions have only even followers. Adding a STAR changes the parity.

8×8 BOARD

	⌄		⌄		⌄		⌄
2ε		2ε		2		$2\uparrow$	
	2ε		$2\varepsilon|2$		$2*$		z_1
2ε		2ε		z_2		z_2	
	2ε		2ε		2ε		2ε
2ε		2ε		2ε		2ε	
	2ε		2ε		2ε		2ε
2ε		2ε		2ε		2ε	

where $z_1 = 2\varepsilon|2\uparrow$ and $z_2 = 2\varepsilon|2*$

7×8 BOARD

⌄		⌄		⌄		⌄	
	2ε		2ε		2ε		2ε
2ε		2ε		2ε		2ε	
	2ε		2ε		2ε		2ε
2ε		2ε		2ε		2ε	
	2ε		2ε		2ε		2ε
2ε		2ε		2ε		2ε	

6×8 BOARD

	⌄		⌄		⌄		⌄
2ε		2ε		2ε		2ε	
	2ε		2ε		2ε		z_1
2ε		2ε		2ε		z_2	
	2ε		2ε		2ε		$\frac{5}{2}*$
2ε		2ε		2ε		z_2	

where $z_1 = 3|\frac{5}{2}*\|2\varepsilon$ and $z_2 = \frac{5}{2}*|2\varepsilon$

Figure 24. Initial Values in Welton's Region.

	✔		✔		✔		✔
	$2|\frac{3}{2}*$		$\frac{3}{2}$		$2*$		$2*$
z_1		$\frac{3}{2}*$		$2*|\frac{3}{2}$		2	
	z_2		$2|\frac{3}{2}*$		$2*$		$2*$
2ε		2		2		2	
	$2\varepsilon|2$		$2*$		$2*$		$2*$
2ε		$2\varepsilon|2*$		$2\varepsilon|2*$		$2\varepsilon|2*$	
	2ε		2ε		2ε		2ε
2ε		2ε		2ε		2ε	

where $z_1 = 2\varepsilon\|2|\frac{3}{2}*$ and $z_2 = 2\varepsilon|\frac{3}{2}*$

Figure 25. 9×8 Values.

Early Values

Like the scrimmage sequence, the early sequence also has a high region and a Welton region. At the lower part of the high region, which is accessible by computer calculations, the early values are observed to be related to the scrimmage values by the equation

$$\mathbf{e}[n] = \mathbf{s}[n - 5]*$$

where n is the altitude of the position $\mathbf{p}$. If the game begins from a conventional initial position with the Fox located near the bottom of a very tall board, we assert that all early values are given by this formula. There is an intermediate region where the Fox is closer to the geese than these early positions, but further away than the scrimmage positions. We have found that almost all such positions have values which can be conveniently expressed as

$$\mathbf{v}(\mathbf{p}) = \mathbf{b}[n - d(\mathbf{p})] \quad \text{if } d \text{ is even}$$

or

$$\mathbf{v}(\mathbf{p}) = \mathbf{b}[n - d(\mathbf{p})] * \quad \text{if } d \text{ is odd.}$$

Here $\mathbf{b}[n]$ is the *backbone* sequence of values. For $n > 31$, we define it as

$$\mathbf{b}[n] = \mathbf{s}[n]$$

although we will subsequently define it slightly differently at certain lower values of n. The integer d, called the *altitude decrement*, depends on local properties of the position, independent of size of the *backfield* between the animal on the lowest occupied rank and the bottom of the board. The values of d for all sufficiently high initial positions appear in the F4 Altitude decrement table shown in Fig. 27.

$\frac{3}{2}*$		$\frac{3}{2}*$		$\frac{5}{4}$		$\frac{3}{2}\mid\frac{5}{4}*$	
	$\frac{3}{2}$		$\frac{3}{2}*\mid\frac{5}{4}$		$\frac{5}{4}*$		z_1
$\frac{3}{2}*$		$\frac{3}{2}*$		$\frac{3}{2}\mid\frac{5}{4}*$		z_2	
	$\frac{3}{2}$		$\frac{3}{2}$		$\frac{3}{2}$		$2*$
$\frac{3}{2}*$		$\frac{3}{2}*$		$\frac{3}{2}*$		$2*\mid\frac{3}{2}$	
	$2\mid\frac{3}{2}*$		$2\mid\frac{3}{2}*$		$2\mid\frac{3}{2}*$		$2*$
2		2		2		2	
	$2*$		$2*$		$2*$		$2*$
$2\!\uparrow$		$2\!\uparrow$		$2\!\uparrow$		$2\!\uparrow$	

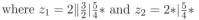

where $z_1 = 2\|\frac{3}{2}\mid\frac{5}{4}*$ and $z_2 = 2*\mid\frac{5}{4}*$

Figure 26. 10×8 Values

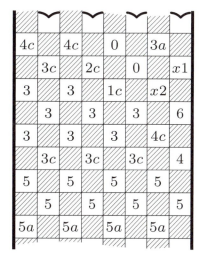

$4c$		$4c$		0		$3a$	
	$3c$		$2c$		0		$x1$
3		3		$1c$		$x2$	
	3		3		3		6
3		3		3		$4c$	
	$3c$		$3c$		$3c$		4
5		5		5		5	
	5		5		5		5
$5a$		$5a$		$5a$		$5a$	

Figure 27. F4 Altitude Decrements.

Of course, Fig. 27 can be reflected into the F0 formation.

At sufficiently high altitudes, the *abc*'s in this table can be ignored; only the numbers matter. As indicated above, these numbers tend to range from 0 at the scrimmage positions to 5 at the early positions when the Fox is still far enough below the geese. Since, after normalizing the parity by adding * to odd positions, the backbone values decrease with large and increasing altitude, n, the Fox favors low values of d; the geese, high values. Evidently a smart Fox will be reluctant to move to the square numbered 6. Each square labelled with the letter x has exceptional asymptotic values which do not occur in the backbone sequence. These tend to be locations a wise Fox will avoid. The special rules for handling these unstable exceptions will be presented in a subsequent section. But first we look more closely at the initial positions at elevations between 31 and 15, a range which is very common in popular play on the 8×8 checkerboard.

The *abc*'s to Trifurcate the Two-Ish Transition

Within the high region of altitudes 32 and higher, the backbone sequence is defined to be equal to the scrimmage sequence, and they give the values of all circled positions of formations A3, A2, A1, B0, F0, and their reflexions. However, at altitude 29, these formations have different values: A31 = F30 = 2ε, but B30 = 2. So A31 and F30 clearly belong to Welton's region; yet B30 clearly does not. Since it violates the parity principle, B30 cannot be considered part of the high region either. So it is not unreasonable to regard altitudes 31, 30 and 29 as the *transitional region* between the high region and Welton's region. In the transitional region, we define the backbone sequence to include this trifurcation (See Table 6).

We can then use Fig. 27 to compute values of lower positions. If the adjusted altitude, $n(\mathbf{p}) - d(\mathbf{p})$, falls in the transitional range, then we use the letter in Fig. 27 to pick the correct

	31	30	29
a	$2\!\uparrow\!*$	$2\!\Uparrow$	$2\varepsilon\|2\!\Uparrow$
b	$2\varepsilon\|2\!\Uparrow\!*\|\|2$	$2\varepsilon\|2\!\uparrow\!*$	
c	$2\varepsilon\|2$	2ε	2ε

Table 6. Truncating the Two-ish Transition.

value of **s**. Of course, we must still adjust its parity by adding a star or not according to the parity of $d(\mathbf{p})$. Omar may verify that, with these conventions, the values obtained from Fig. 27 match those we have reported from Siegel's *cgsuite* whenever the translated altitude lies in the transitional region.

Extending the Backbone to Lower Altitudes

Below the transition region, we define the backbone values as

$$\mathbf{b}[n] = 2\varepsilon \text{ for } 28 \geq n \geq 17.$$

This is Welton's region. Below Welton's region, we define two more backbone values as

$$\mathbf{b}16 = \tfrac{5}{2}*|\mathbf{2over} \text{ and } \mathbf{b}15 = \tfrac{5}{2}* \,.$$

Omar can now verify that all of the numbered entries in Fig. 27 yield the same initial values as we have reported from Siegel's *cgsuite* for boards of sizes 7×8 through 10×8. The values it gives for the 6×8 board are not inconsistent, although it fails to define the values for a few exceptional locations on this board.

Although the decrements we have contrived in Fig. 27 and the backbone sequence give all the correct values of all initial positions on boards 7×8 or taller, many of these contrived decrements are not unique. They should be regarded primarily as a shorthand way of summarizing the much larger quantity of data needed if each translation of the F4 formation is treated separately.

For formations of height greater than 36, none of the abc's has any relevance, because then the altitude, which is the height minus the decrement, always exceeds 31, the largest altitude in the trifurcated region. So beneath the lowest row shown in Fig. 27, all locations are occupied by an unlettered 5.

Early Values of Other Formations

Fig. 28 shows the decrement tables for several other common formations. Like F4, all of these tables also apply to the reflected positions. And like F4, all of these tables also have only unlettered 5s below the lowest row shown. Squares left blank all have very hot values, with **off** as a Right follower. From Welton's region on up through the highest position that can fit

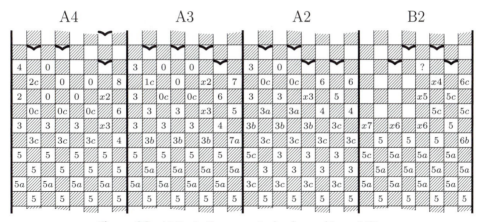

Figure 28. Altitude Decrements for Some A's and B's.

onto a 14×8 board, every value obtainable from any of these tables agrees with those obtained from Siegel's *cgsuite*, and we conjecture they remain valid at arbitrarily large heights.

Evidently, when wise geese can choose between a pair of formations such as A2 and B2, they do best to take the Fox's altitude into account, even when he is sufficiently distant from the geese that escaping is not an imminent threat. Often a "just-in-time" defense yields a higher value for the geese than a prematurely "solid" position.

It is convenient to define the "separation" of a position as the minimum gooserank minus the Foxrank. When the separation exceeds 2, the M formations are the best we have been able to find. As wise geese have known for ages, the fastest and safest ways to move southward with minimal effort is to rotate around between M4 and M0. How often does the lead goose change?

Migrating geese can play with their eyes closed as long as the separation exceeds 2. Then they may need to concentrate on winning whatever escapade(s) the Fox attempts, such as

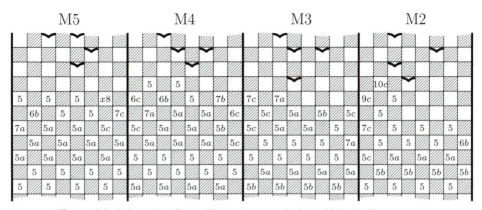

Figure 29. Migrating Geese Formations and their Altitude Decrements.

Escapade 6 on page 680. Once the Fox is this close, temptations to make suboptimal moves (c.f., the questionable move in Variation 3′ of Escapade 6) are rare and easily avoided. In our opinion, it is more difficult to learn how to play other geese formations well because they require the geese to make important decisions earlier, when the Fox is further away and the consequences of the next few moves are harder to foresee.

Resolving the Exceptions

Entries beginning with an "x" in these decrement tables correspond to locations slightly hotter than any values which appear in the asymptotic backbone sequence. There is a technical sense in which they are unstable. However, they all flow into the backbone sequence within a very small number of moves, and we use this fact to represent them in the following convenient shorthand notation:

$$
\begin{aligned}
x1 &= \text{``}9c|2a\text{''} \\
x2 &= \text{``}7c|0\text{''} \\
x3 &= \text{``}5c|-1c\text{''} \\
x4 &= \text{``}6c|-1\text{''} \\
x5 &= \text{``}5c||-2|\mathbf{off}\text{''} \\
x6 &= \text{``}3c||0|\mathbf{off}\text{''} \\
x7 &= \text{``}7c||0|\mathbf{off}\text{''} \\
x8 &= \text{``}8c|||3c||0|\mathbf{off}\text{''}
\end{aligned}
$$

So, for example, consider the square marked "$x1$" near the top right corner of Fig 27. On a 10×8 board, its row-rank is 7, two below the geese on rank 9. The height of the position is $4 \times 9 = 36$, and the altitude is $36 + 7 = 43$. Following the shorthand, we write the value of this position as

$$
\mathbf{b}[43-9]\,|\,\mathbf{b}[43-2]*,
$$

(where the * is added to the Right follower to maintain parity)

$$
= \mathbf{b}[34]\,|\,\mathbf{b}[41]* = 2\|\tfrac{3}{2}\|\tfrac{5}{4}*
$$

Since both immediate followers are an odd number of moves from the current position, we must add a star to each of them to maintain parity, in addition to the star which is added to compensate for any odd decrement of the index.

In Welton's region, some of these expressions simplify because of confluence caused by multiple appearances of the value 2**over**. Asymptotically, several of the referenced backbone values are switches. In particular, $x1$ and $x3$ asymptotically represent the same 3-stop; $x2$ and $x4$ both yield a 2-stop hotter than any comparable backbone value; $x5$ and $x6$ yield 3-stops if **off** is regarded as the Right stop of the Right follower; and $x7$ is a 4-stop with the same property.

When a decremented index falls in the trifurcated range, the *abc* letter specifies which value to select, allowing these formulas to remain valid down into Welton's region. In the F4, A and M formations, values and formulas remain valid all the way down to altitude 15.

The B2 formation is unusual in several respects. Not only does it have a relatively high number of locations with exceptional decrements, its higher circled Fox location, marked with "?" in the decrement table, has values which agree with the backbone sequence (decremented by 0) at all altitudes above 31, but has the value of 2 rather than 2ε at altitudes 26 and 22. So this sequence of B2 positions has no Welton region. It is the only such location in any of the decrement tables shown in this book.

We assert that all of the values obtained by applying our decrement tables are valid even at arbitrarily high altitudes. Although this must still be regarded as a conjecture, it is closely coupled to a scrimmage sequence which is solidly anchored (at all altitudes congruent to 2 mod 5) to an infinite number of points for which we earlier gave a complete proof.

Museums

Why Museums Contribute Nothing to Current Production Quotas

Anyone hoping to become a true Fox-and-Geese + Hackenbush expert must surely learn to play competently from the initial geese formation on any size of board, with the Fox at any initial location, and when accompanied by any Hackenbush position. All of the material we have presented in this chapter thus far has been directed towards that objective. Some of the positions we have studied have been *unstable* in the sense that they occur only within a reversible sequence between two *stable* positions. Good players soon learn to avoid spending much of their time worrying about the values of such unstable positions, which have canonical followers too big or too small to have any effect on the more fundamental stable positions between which they appear. Like professional Go players, they soon sense which moves "have sente." They waste no further effort detailing how badly the opponent can be trounced if he fails to respond to such a move.

So, although there is a sense in which, with considerable assistance from Siegel's *cgsuite*, we have "solved" all standard versions of F&G + H, we do not know the value of a totally arbitrary position, in which all four geese as well as the Fox might be placed anywhere on a large board. We have shown that from any conventional starting position, rational players would never reach such a state. Like the Garden of Eden, such a position can exist in our world only if it was created that way before any other history began.

Why Museums Provide So Much Artistic Attraction

Artists and mathematicians often find considerable beauty in rare objects. Our own collection of Fox and Geese museum pieces is presented on the following pages. See for yourself and admire! But we must confess that many of our museum pieces are not genuine originals; they were mere copies taken from a larger museum discovered, collected, and arranged by Aaron Siegel. A weblink to his F&G museum may be found at http://cgsuite.sourceforge.net

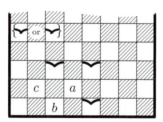

$$a = 1 \text{⇑}*; \quad b = 1 \text{⇑}; \quad c = 1 \text{⇑}*$$

1-ish Positions with High Atomic Weights

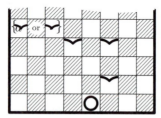

Hot Atomic Weight—Value $= 2 \text{⇑}* \mid 2*$

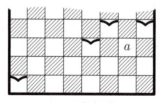

$$a = 0 \| 0 | \textbf{off}$$

The Tiniest Value in any Possible Game: $0\|0|\textbf{off}$

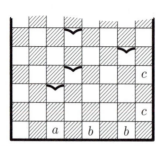

$$a = 4\,\textbf{Tiny}(2*); \quad b = 2\{\varepsilon|0\}; \quad c = 2\{\varepsilon|*\}$$

Tiny and Switches of Temperature ε

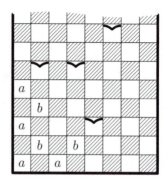

$$a = \uparrow; \quad b = *$$

Pure Infinitesimals

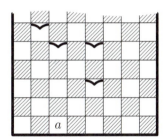

D17 Has a Very Complicated Value (See below).

$$a = \left\{ 6* \,|\, 2\mathbf{over} \right\}, \left\{ \{5* \,|\, 4\}, 4 \,\Big|\, \{4 \,|\, \tfrac{7}{2} \,\|\!|\, 2\,\text{⛫}\, \|\, 1 \,|\, \mathbf{off}\}, \{4 \,\|\!|\, 2\,\text{⛫}\, \|\, 1 \,|\, \mathbf{off}\} \right\} \,\Big|\, b$$
$$ \;\; d\,c \quad e \qquad\quad \beta\,\alpha\gamma \quad k\,f \quad zxy \;\; g \quad \delta \;\; i \;\; j \qquad \gamma\;h \quad \delta \;\; i \;\; j \qquad a$$

$$b = \left\{ 4 \,\|\!|\, 2\,\text{⛫}\, \|\, 1 \,|\, \mathbf{off} \right\}, \left\{ \{4* \,|\, 3, \{4 \,|\, \tfrac{7}{2} \,\|\, 1\}\}, \{4* \,\|\!|\, 4 \,|\, \tfrac{7}{2} \,\|\, \mathbf{off}\} \,\|\!\|\, 2 \,|\, 2, \{2 \,|\, \mathbf{off}\} \,\|\, \tfrac{3}{2} \right\} \,\Big|\, \mathbf{off}$$
$$ \;\; \gamma\;h \quad \delta \;\; i \;\; j \qquad\quad r\;q \quad zxy\;s\;t \qquad w\;u\;zxy\;v \qquad m \quad p \qquad\quad n\,o \quad b$$

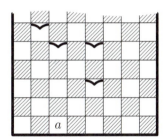

$a, b, c, d, e \qquad\qquad f, g, h, i, j, k \qquad\qquad m, n, o, p$

$q, r, s, t \qquad\qquad u, v, w \qquad\qquad x, y, z$

$\alpha, \beta, \gamma \qquad\qquad \delta, \epsilon, \ldots$

Some Canonical Positions of a.

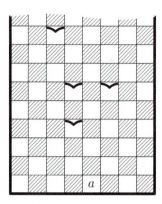

The Most Complicated Value Known.

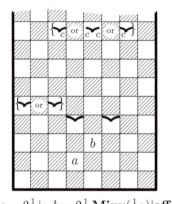

$$a = 2\tfrac{1}{2}\!\downarrow;\ \ b = 2\tfrac{1}{2}\,\mathbf{Miny}_{c}(\tfrac{1}{2}*)|\mathbf{off}$$

Negative Infinitesimals.

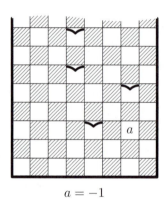

$$a = -1$$

Negative Number: -1.

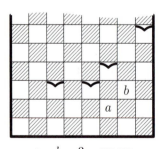

$$a = b = 3 - \mathbf{upon}$$

Loopy Infinitesimal of Higher Order than ε.

Solutions to Problems

1. Fox-Flocks-Fox. In view of the values of the initial positions on the 4×8 board in Fig. 23, second player can win in all-but-one of the 12 possible symmetric pairs of initial Fox positions.

2. Escapade 6 is the Migrating Geese Formation M3. If Geese move first, they must refrain from Migrating to Formation M2, because the Fox is too close. Instead, they can begin by moving the eastern goose southeast, giving them a one-move advantage in all variations of Escapade 6.

3. Solutions of Escapades 7 and 8 appear in Figures 30 and 31.

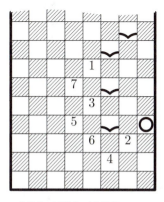

8@4, 10@2, 12@O, etc.

Figure 30. Solution to Escapade 7: Fox Escapes from Opening of 5×2 Cage. But without Bottom Row, Geese Could Play 5@6a to Win More Easily.

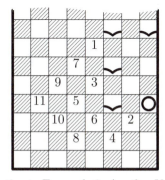 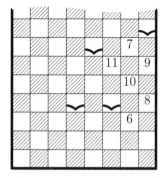

Figure 31. Solution to Escapade 8: Another Variation. Clever Geese Thwart Fox's Attempt to Escape from Entrance to his 4×2 Cage.

Open Problems

1. Define a position's *span* as the maximum occupied row-rank minus its minimum occupied row-rank. Then quantify and prove an assertion such as the following: If the backfield is sufficiently large, and if the span is sufficiently large, and if the separation is sufficiently small, and if the Fox is niether already trapped in a daggered position along the side of the board, nor immediately about to be so trapped, then the Fox can escape and the value is **off**.

2. Show that any formation of three geese near the centre of a very tall board has a "critical rank" with the following property: If the northern goose is far above, and the Fox is far below, then the value of the position is either positive, HOT, or **off**, accordingly as the northern goose is closer, equidistant, or further from the critical rank than the Fox.

3. Welton asks what happens if the Fox is empowered to retreat like a bishop, going back several squares at a time in a straight line? More generally, suppose his straightline retreating moves are confined to some specified set of sizes. Does {1,3}, which maintains parity, give him more or less advantage than {1,2}?

4. What happens if the number of geese and boardwidths are changed?

Maharajah and Sepoys

As we said in the Extras to Chapter 19, there are very many games involving encirclement, often mixed with various forms of capture. The name Fox and Geese, for example, has also been used for various games played on the English Solitaire board (Chapter 23) a couple of which are described in the Extras to Chapter 19. Most of these games are typified by a considerable numerical imbalance between the opposing forces. This is compensated by much greater mobility of the numerically inferior pieces. An extreme example is Maharajah and Sepoys which is played like ordinary Chess. White has a standard set of 16 pieces, starting from their usual positions, while Black has a single piece, the Maharajah, who starts on any unoccupied square and can move like a Chess Queen *or* a Chess Knight. The object of both sides is checkmate, of either the White King or the Maharajah. As in most of these games the Lord is on the side of the big battalions, and White wins with correct play.

References and Further Reading

Robert Charles Bell, *Board and Table Games from Many Civilizations*, Oxford University Press, London, 1969.

Elwyn Berlekamp, Idempotents Among Partisan Games in *More Games of No Chance*, Richard Nowakowski, ed; vol 42 in MSRI Publications, Cambridge University Press, 2002.

Maurice Kraitchik, *Mathematical Recreations*, George Allen and Unwin, London, 1943.

Fred. Schuh, *The Master Book of Mathematical Recreations*, transl. F. Göbel, ed. T. H. O'Beirne, Dover. N.Y., 1968. Chapter X: Some Games of Encirclement, pp. 214–244.

David Wolfe, The Gamesman's Toolkit in *More Games of No Chance*, Richard Nowakowski, ed; pp. 93–98, MSRI Publications 42, Cambridge University Press, Cambridge, UK, 2002.

-21-

Hare and Hounds

I like the hunting of the hare
Better than that of the fox.

Wilfred Scawen Blunt, *The Old Squire*.

The French Military Hunt

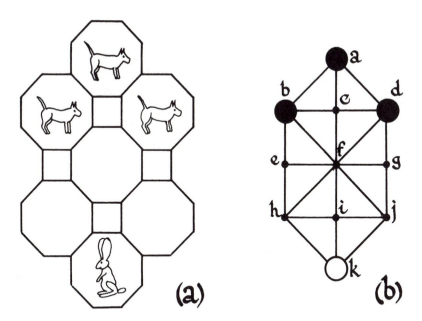

Figure 1. The French Military Hunt. Hare and Hounds on the Small Board.

This little game is very like Fox and Geese. It features a hunter whose three hounds (dogs) try to trap a hare (rabbit) on the board shown in Fig. 1(a). If you can't persuade enough animals to make the right manoeuvres, you can play with four coins on the nodes of the equivalent board shown in Fig. 1(b). It becomes more interesting on the larger board of Fig. 2. At each

711

turn the hunter moves any one hound to a neighboring empty place, and the hare makes a similar move. However the hounds, starting from the top, may not retreat, although a hound may go back and forth horizontally as between e and f in Fig. 1(b). The hare is completely free to advance or retreat or move horizontally. The hounds win by trapping the hare so that he cannot move at his turn. If the hounds fail to advance in ten consecutive moves, the game is usually declared a win for the hare.

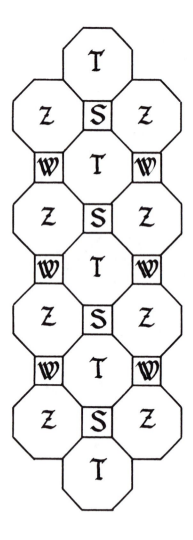

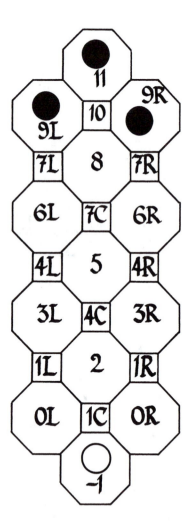

Figure 2. The Larger Board, with Four Types of Place.

Figure 3. The Larger Board, Numbered for the Trace.

Two Trial Games

If you want to see how the game goes, first set up the board and watch an expert hunter against a novice hare:

hounds: *abd*   *cbd*   *fbd*   *fed*   *fhd*   *fhg*   *fhj*   *ihj* (wins)
hare: *k*   *i*   *j*   *g*   *j*   *i*   *k*

First game.

The chase looks so easy that the novice decides to direct the hounds in pursuit of an expert hare:

hounds: *abd*   *cbd*   *fbd*   *fed*   *feg*   *fhg*   *fig*   *eig*   *fig*   *fij*
hare: *k*   *j*   *i*   *h*   *k*   *j*   *k*   *j*   *k*   *h*

Second game.

and now the hare will escape by *e* or *f*.

If expert hounds chase an expert hare on Fig. 1, who wins? And what if the hare makes the first move? Or starts from a different place? (See the Extras.) And (when you've become more expert) what about Fig. 3?

History

According to Lucas the game (on Fig. 1) was popular among French military officers in the nineteenth century. Some say it was invented by Louis Dyen; others attribute it to Constant Roy. It was solved by Lucas (1893) and Schuh (1943) and popularized (again) by Martin Gardner (1963). Schuh's analysis was based on a list of 18 classes of winning positions for the hounds (reproduced in the Extras) and he recognized that "the opposition" plays a key role, but he had no exact definition for it. In a later section we'll give a definition which simplifies the game on Fig. 1 and also allows us to solve that on Fig. 3.

The Different Kinds of Place

Let's look at the board more closely. There are really two types of octagon: central ones (T in Fig. 2) and side ones (Z). There are also two types of square: central squares (S) and side squares (W). Except near the very top or bottom of the figure, each T or Z is next to at least one place of every other type, but each W or S is only next to octagons, T and Z. Since W and S are never adjacent, it's sometimes convenient to lump them together into a single class, N. Of the three types T, Z and N, every place, even the ones at the top and bottom, is next to at least one place of each other type, but to none of its own type. The letters correspond to remainders after division by 3 of the numbers from Fig. 3:

| Remainder Zero : Z |
| Remainder oNe : N = Weak or Strong |
| Remainder Two : T |

In Fig. 3 the difference of two numbers in adjacent places is always 1 or 2.

The sum of the numbers occupied by the four animals is an important property of the position; we call it the **trace**. Every move changes the trace by 1 or 2. If the hounds succeed in trapping the hare at the bottom of the board, then the hounds are at 0, 1, 0 against the trapped hare at −1 and the trace is 0. If instead the hounds trap the hare on the side of the board, say at 1L, then the hounds end on 3, 2, 0 against the hare on 1, and the trace is 6. It can easily be checked that

> No matter where
> You trap the hare,
> The trace you'll see
> Divides by *three*

TRIALITY TRAPS!

The Opposition

The best way for the hounds to make their trap is to move so that they leave the trace a multiple of 3 at every turn. We call this "keeping the opposition". If they do this, the hare's move must be to a non-multiple of 3, because it changes the trace by 1 or 2. But whenever the trace is not divisible by 3 the hunter usually has a choice of several hound moves which restore it to a multiple of 3, and among these he should find one that restores a winning position.

> Threefold traces
> Win most chases.

KEEPING THE OPPOSITION

If you check the traces for our first game, with the board numbered as in Fig. 4, you'll see that the hounds always kept the opposition:

hare:	k	i	j	g	j	i	k
hounds:	cbd	fbd	fed	fhd	fhg	fhj	ihj
trace:	9	9	6	6	3	3	0

Since the hare doesn't want to be trapped, he doesn't want the hunter to move to positions whose trace is divisible by 3. The best way to prevent this is for the hare to grab the opposition by moving to such a position himself. Then any hound move will change the trace to a nonmultiple of 3 and the hare is likely to be able to regrab the opposition. This is the way the hare won our second game. The hounds blundered on their second move by playing from 4 to 2, giving a trace of 8, and from then on the hare managed to retain the opposition at every turn:

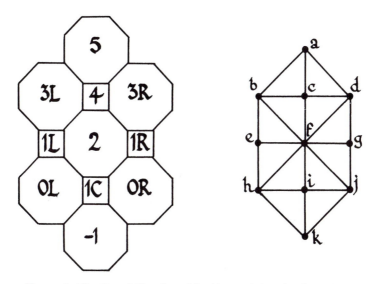

Figure 4. The Board Numbered for Determining the Opposition.

hounds:	abd	cbd	fbd?	fed	feg	fhg	fig	eig	fig	fij
hare:	k	j	i!	h	k	j	k	j	k	h
trace:	10	10	9	6	3	3	3	3	3	3

So whoever can move to a position whose trace divides by 3 is said to have the opposition. The opposition is certainly a valuable commodity which both players desire. But it's not all there is to the game, because sometimes the *hounds* may have the opposition but be unable to keep it without letting the hare escape behind them. In other cases the *hare* may have the opposition for several moves, but then lose it because the hounds block his only moves to places which would restore it. However, such positions are rather rare, and the average player who combines the principle with a little commonsense will usually trap a novice hare on the small board. An annotated example appears on p. 716.

When Has the Hare Escaped?

He has **escaped** if he has passed or is passing two hounds, unless he is on a *square* place (W or S) and the hounds can immediately occupy the neighboring octagons (Z or T) aside or ahead of him.

Although he may have not escaped, the hare is **free** in some other positions in which the hounds can never force him to retreat. This certainly happens if he's strictly passed a hound and is not on a Weak (W) square, or, if he's on a central octagon (T) and is past or passing at least one hound.

Third Game

Hounds	Hare	Trace	Comments
3L, 5, 3R	−1	10	
3L, 4, 3R		9	Taking the opposition
	0R	10	
2, 4, 3R		9	{ A novice hunter might have moved 4 to 2, giving a "solid" position, but losing the opposition.
	1C	10	
2, 3L, 3R		9	{ The other "reasonable" move, 3R to 1R, changes the trace by the wrong amount. Since the move from 2 to 1 would allow the hare to escape, there's really only one choice.
	−1	7	
1C, 3L, 3R(!)		6	{ Because the hounds can't retreat, they can never increase the trace by 2, so to gain the opposition they must decrease 7 to 6 by moving a hound from 2 to 1. A move to 1R or 1L won't lose, but wastes time, since the hare can force the hounds back to the present position position by going to 1C.
	0R	7	
1C, 2, 3R		6	{ The other two moves (3R to 2, 1C to 0L) that restore the trace to 6 would let the hare escape.
	−1	5	
1C, 2, 4(!)		6	{ Once again, the other moves (3 to 1, 2 to 0) keeping the opposition would let the hare escape, leaving only this unlikely looking move.
	0R	7	
0L, 2, 4(!)		6	{ 4 to 3R repeats; 2 to 1 allows escape; only 1C to 0L makes progress
	1R	7	
0L, 2, 3R		6 ⎤	
	0R	5 ⎬	Obvious
0L, 2, 1R		3 ⎦	
	−1	2	Hare's last gasp.
0L, 0R, 1R		0	The novice hunter might now lose by playing from 1R to 0R.
	1C	2	
0L, 0R, 2		3	{ The only time the hounds reach a trace larger than their previous one.
	−1	1	
0L. 0R. 1C		0	Wins.

Losing the Opposition

To analyze the exceptional positions, when someone wins in spite of not having the opposition, it's best to consider the types of place the animals occupy. For example, all the positions where the hounds have just won are of type Z^2NT, meaning that 2 animals are on Z places, 1 on N and 1 on T.

Some of the exceptional cases arise from the difference between the Strong and the Weak types of N places. Each Strong (central) N square is next to *four* other places, while each Weak (side) square is next to only three. Other things being equal, an animal should prefer a Strong place to a Weak one, since both make the same contribution to the opposition; but the Strong place is likely to offer him more choices later. For example, one exceptional case arises when the hounds move to Fig. 5. Despite the fact that the hounds have the opposition

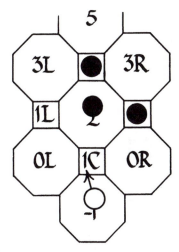

Figure 5. An Exceptional Hare and Hounds Position.

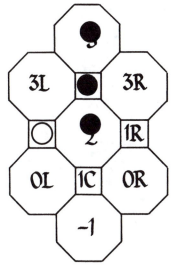

Figure 6. Another Exception to the Opposition Principle.

(trace 3), the hare wins by playing to 1C, because now the only hound moves which keep the opposition let the hare escape. In some sense this N^2T^2 position loses because the hound at 1R is on a Weak square. On the other hand, we saw in our third game that a hare on -1 has no defence against hounds on 4C, 2, 1C (another N^2T^2 position). Unless the hare has passed one or more hounds, S^2T^2 wins for the hounds, but SWT^2 often loses.

As another example, suppose the *hare* has just moved to the position of Fig. 6. He has the opposition, but after the hound on 4C moves to 3L, the hare must retreat to 0L, losing the opposition and the game. But a hare in place 1C against these hounds would have both the opposition and a winning position. Once again, the difference between a Strong and a Weak square means the difference between winning and losing, this time for the hare.

A Strategy for the Hare

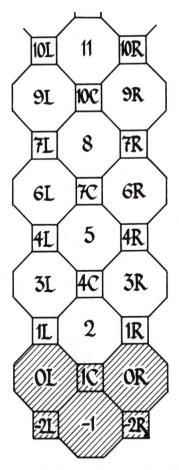

Figure 7. Keeping the Opposition on a Semi-infinite Board.

We'll show that an expert hare that has the opposition on the semi-infinite board of Fig. 7 can either keep it indefinitely or escape, unless he has to start from the **Scare'm Hare'm** position (Fig. 8). In fact the hare will always stay on the six shaded places numbered 1C, 0L, 0R, −1, −2L and −2R, unless the hounds let him out. His basic strategy is to keep the opposition.

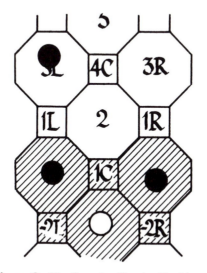

Figure 8. The Scare'm Hare'm Position.

> If possible, escape or gain your freedom!
> Otherwise, keep on the six shaded places, and
> if you can keep the opposition by a move to a
> non-Weak place, do so.
> If a move to S(1C) is blocked, then
> (A) against hounds on T^2S, move to W(−2L or −2R),
> (B) against hounds on ZN^2, advance to Z (losing the
> opposition) on the other side of the
> board from the hound-occupied Z.
> If a move to Z (0) is blocked,
> (C) go to T (−1) (losing the opposition).

THE HARE'S STRATEGY

If these rules allow two or more moves, choose any one. If they allow none, resign (or hope for a mistake)!

First we show that if the hounds reach Fig. 8, a recent hare's move must have been of type (A), (B) or (C). For if the hounds came from a position in which they *had* the opposition,

then the hare, after his last move, *didn't*, and the present position must have been reached by
(B) or (C). Otherwise the hounds have come from a position whose trace was congruent to 1,
mod 3, and hence from Z^2N, since they are on Z^3 in the figure. At the hare's last move 2 was
vacant and either 0L or 0R was occupied by a hound. But if the hare came from 0L, 1C or
0R he could have escaped by moving to 2 and so he must have come, from the Weak square
-2, which he can only have reached by a move of type (A).

Suppose you've just made a move of this strategy which was not of type (A), (B) *or* (C).
Then you *have* the opposition and you're *not* on a Weak square and the table below shows
that the Hare's Strategy always gives you another move, unless you're faced with the Scare'm
Hare'm Position.

To From	Z	S	T
Z	—	(A) or gain freedom by advance to T.	already free
S	escape by advance to T.	—	already free.
T	escape, since *not* Fig. 8.	(A) or (B)	—

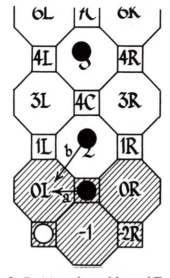

Figure 9. Position after a Move of Type (A).

Next suppose you've just made a move of type (A). Then in the next few moves you can either escape or regain the opposition by a move *not* of type (A), (B) or (C), and from which the hounds can't immediately move to Fig. 8. This is because when (A) is applied, the Strong square 1C must be occupied and also two central octagons (not including −1 because the hare has not escaped); see Fig. 9. Now the only way the Hare's Strategy can lose the opposition from an N square is by a move of type (C) after a hound moves to 0L. But after move (a) in Fig. 9 the hare regains the opposition, while after move (b) he soon escapes. The hounds can't reach the Scare'm Hare'm Position in time.

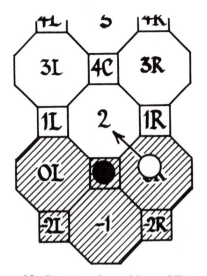

Figure 10. Position after a Move of Type (B).

Now suppose you've just made a move of type (B) (Fig. 10). Then you threaten to escape by moving to the empty T place ahead of you. If the hounds fill this from N, you escape by advancing to W, and if a hound from Z fills it you can reacquire the opposition by retreating to T. The hounds can't straight away reach the Scare'm Hare'm Position.

Finally, *if you've just made a move of type* (C), and were on a Strong square, *both* adjacent Z's must be occupied and you could have escaped. So you were on a Weak square and we have already discussed the situation following your previous move, which must have been of type (A).

On the Small Board

The Hare's Strategy shows that if they don't have the opposition the hounds can only win on the small board by keeping a hound on 5 until they can grab the opposition by moving him to 4 or 3. If they move first from 3L, 5, 3R the hounds can beat a hare starting anywhere except 4. Here is a sample game.

Hounds	Hare	Remarks
3L, 5, 3R	1C	(Or the hare could start on 1L or 1R.)
3L, 5, 2		
	−1	If instead to 0, the hounds take the opposition by moving from 5 to 4.
1L, 5, 2		
	1C	If instead to 0, the hounds take the opposition by moving from 5 to 3.
0L, 5, 2		
	−1	If instead to 0, the hounds take the opposition by moving from 5 to 4.
1C, 5, 2		
		Now, since there is no place −3 on this board, the hare is forced to give the hounds the opposition and the game.

On the Medium and Larger Boards

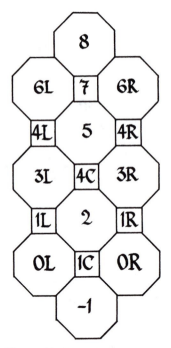

Figure 11. The Medium Board.

By a slight extension of this argument, the hounds, moving from 6L, 8, 6R on the Medium Board (Fig. 11) can trap a hare starting on −1, 0, 2, 3 or 5. Since they have the opposition they can certainly win on the Small Board got by dropping numbers −1, 0 and 1 (Fig. 1). The hare on 2 may reach one of the positions of Fig. 12, forcing the hounds to give him the opposition in return for his retreat, but it is too late, since the hounds can play to 3L, 5, 3R, which wins for them, even without the opposition, because places numbered −2 are not on the board. What if the hare now goes to 0L? See the Extras.

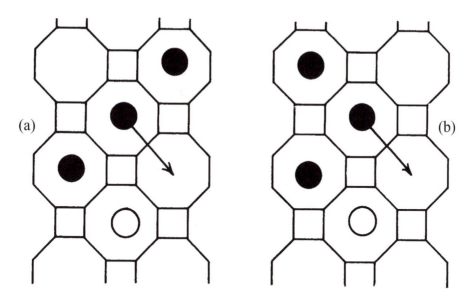

Figure 12. A Sound Bound for a Hound?

It is interesting that the hounds win if the configuration of Fig. 12(a) occurs at 3L, 5, 6R against 2, but not if it is higher (on the Larger Board of Fig. 3) at 6L, 8, 9R against 5. After the hound moves from 8 to 6, the Hare snatches the opposition by retreating to 3 and then follows his Strategy, but using the next set of six squares (4C, 3L, 3R, 2, 1L and 1R) up the board.

It should now be clear that the Hare's Strategy can be improved. If the Hare doesn't have the opposition, he should try to reach a position like 5 against hounds on 6L, 8, 9R (all such positions have trace 28). The way to force the hounds to move into such a position is to move to one whose trace is larger than the desired one by a small multiple of 3. In fact we can prove that

> on the Larger Board (Fig. 3) the
> hounds can win from a position
> of trace 31 only if the Hare is on a
> Weak square or the position is
> 6, 10, 11 *versus* 4C (Fig. 13).

THE THIRTY-ONE THEOREM

The proof is sketched in the Extras.

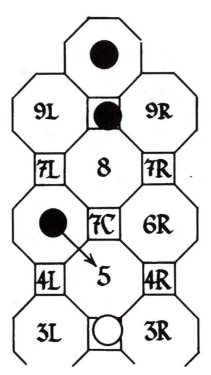

Figure 13. The Hound-Dog Position.

Extras

Answers to Questions

Against the hounds placed as in Fig. 1, the hare can win only if he starts at c and requires the hounds to start first.

With the hounds on 9L, 11, 9R in Fig. 3 the hare can win from any position provided he has first move. The hardest case is when he starts on 1C, so that the hounds have the opposition. He wins by playing to 2 and using the Thirty-one Theorem. Of course if he *starts* on 2 he plays to the Strong square 4C *and* gains the opposition.

If the hounds move first they can win *only* if the hare starts on −1. They must play with great care, not only maintaining the opposition but also preventing the hare from escaping or achieving the trace 31. Surprisingly, even though it gains the opposition, the opening move from 11 to 10 loses! The difficulty is that if the hare advances via 0, 1 and 3 to 5, the hounds must then be able to reach 6L, 10, 6R. The defence 6L, 7, 6R is unattainable against a hare who is determined to keep the trace at least as high as 27. The defence 6L, 7R, 9R fails when the hare moves from 5 to 7C, forcing the 7R hound to occupy 8, and then retreats to 5 again and wins as in Fig. 12. If the hounds try to prevent the hare from reaching 5 by occupying 5, 9, 10, say, when the hare is on 3, then he escapes via a weak 4. But how can the hounds reach 6L, 10, 6R if the hare plays via 0, 1 and 3 to 5? They must have come from 6, 8, 10 if they had the opposition with the hare on 3; but where were they before that with the hare on 1? There is no position leading to 6, 8, 10 in which they had the opposition!

A Sound Bound for a Hound?

If the hare is 0L and the hounds are on 3L, 5, 3R, how do they win? Answer 3R to 2. If hare takes the opposition by going to −1 then 2 to 0R, and if hare to 1C, then 3L to 2. If hare to −1 again, 0R to 1C wins; the trick is to hold back the hound on 5 until they're ready for the kill.

All Is Found for the Small Board Hound

In this chapter we've normally taken the point of view of the hare. To redress the balance, Figs. 14 and 15, which are adapted from Figs. 92 and 93 on pp. 241, 243 of Fred. Schuh's Master Book of Mathematical Recreations, show all the winning positions for the hounds on the Small Board. Figure 14 is a minimal set of 24 $\mathcal{P}$-positions (hounds win if hare has to move) which will ensure victory in all the positions the hounds deserve to win. Figure 15 shows 13 other $\mathcal{P}$-positions for the hounds which they can use for variety in seeking to hide their strategy from inquisitive hares. In each of the 37 positions the hare's place is indicated by a number, the remoteness function for the position.

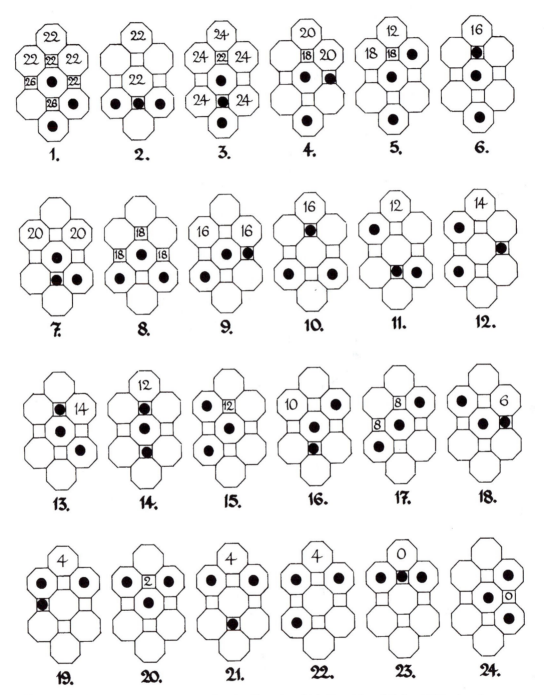

Figure 14. Twenty-four $\mathcal{P}$-positions for the Hounds which Provide a Minimal Winning Strategy.

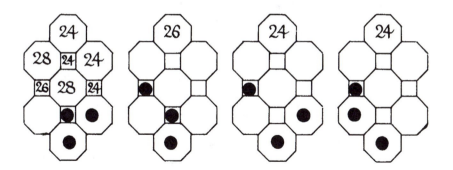

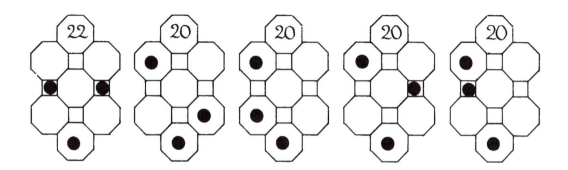

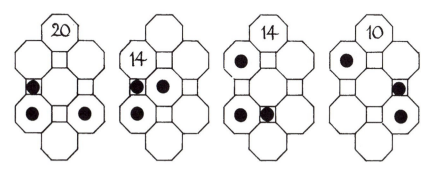

Figure 15. Thirteen More $\mathcal{P}$-positions for the Hounds.

Table 1 gives a winning strategy for the hounds, based on the first 20 $\mathcal{P}$-positions of Fig. 14. The remarks are listed here (L and R are *hare's*; in Figs. 14, 15 left and right are the *hounds'*):

(a) the hounds do not have the opposition, but the hare is now forced to 1R, whereupon the hounds go to 3R, 5, 2 (position 1, reflected), still without the opposition, but the hare is forced again. After he goes to 0R, the hounds to go 3R, 4, 2 (position 7, reflected).

(b) also without the opposition, but see position 4.

(c) still without the opposition, but see position 5.

(d) even now the hounds don't have the opposition, but (position 6) the hare is forced to a zero place and the hounds go to position 13 or its reflexion.

(e) if now or later the hare goes to -1, play as from position 18: the hound on 2 goes to 0, (position 21 or 22), and then the rear hound comes to 2 (position 20).

From:	if hare plays on:	hounds reply by moving to:	arriving at:	with remoteness:	and trace:	Remarks
initial	4,1R,1C,0,-1	3L 5 2	1.	26,26,22,22,22	14,11,11,10,9	
position	2	3 4 3	2.	22	12	
1.	3R	4 5 2	3.	24	14	(a)
	-1	1L 5 2	4.	20	7	(b)
1. or 2.	0	3L 4 2	7.	20	9	
	1	3 2 3	8.	18	9	
4.	1C!	0L 5 2	5.	18	8	(c)
	0	1L 3R 2	9.	16	6	
5.	-1!	1C 5 2	6.	16	7	(d)
	0R	0 4 2	16.	10	6	
6.	0L	1C 3L 2	13.	14	6	
7.	1!	3 3 2	8.	18	9	
	-1	3L 4 0R	11.	12	6	
8.	0	1L 2 3R	9.	16	6	
	-1	3 1C 3	10.	16	6	
9.	-1!	1L 3R 0R!	12.	14	3	
	1	0L 2 3R	17.	8	6	
10.	0L	3L 1C 2	13.	14	6	
11.	0L	2 4 0R	16. refl.	10	6	
	1C	3L 2 0R	17. refl.	8	6	
12.	1C	2 3R 0R	15.	12	6	
	0L	1L 2 0R	18.	6	3	
13.	-1	4 1C 2	14.	12	6	
	1L?	3L 0L 2	trapped!	0	6	
14. or 15.	0L	4 0R 2	16. refl.	10	6	(e)
16.	1	3R 0L 2	17.	8	6	(e)
17.	0R	1R 0L 2	18. refl.	6	3	(e)
18.	-1	1L 0 0	19.	4	0	
	1C	0 0 2	20.	2	3	
19.	1C	2 0 0	20.	2	3	
20.	-1	1C 0 0	trapped!	0	0	

Table 1. A Winning Strategy for the Hounds Using just the Positions of Figure 14.

Proof of the Thirty-One Theorem

The hounds can only keep the opposition from a position of trace 31 by moving to 30 (a move to 33 would involve an illegal retreat). If they go to 30, the hare will move to 31 if he can; the hounds will move back to 30 and the hare will win by repetition. The hounds can only win by getting the hare on to a Weak square, or by preventing him from moving to 31. How might they do that? If the hare is on r, the hounds must be blocking any strong neighboring place $r + 1$. Suppose the other hounds are on x and y where $x + y \leq 11 + 10$.

$$r + (r + 1) + x + y = 30,$$

$2r + 1 \geq 9$, $r \geq 4$. If $r \geq 8$, $x + y \leq 13$ and the hare has escaped.

If $r = 7$, a hound must be blocking 8, $x + y = 15$, and, unless the hare has escaped, $x = 6$, $y = 9$. The hare moves to 5 and reaches Fig. 12.

If $r = 6$, a Z place, the hounds must be blocking 7C and also $x = 8$, to prevent escape, so $y = 9$. The hare plays to 5. If the hounds restore the trace to 30 the hare returns to 6 and wins by repetition.

If $r = 5$, a T place, the hounds must be blocking 6L, 6R and $y = 13$, off the board.

If $r = 4$ and not a Weak square, the hare is on 4C. A hound must be blocking 5, so $x + y = 21$, $x = 10$, $y = 11$. This is the exceptional Hound-Dog Position (Fig. 13) which the hare can't win. If he goes to 3, the hound on 10 moves to 8. If the hare then moves to 4R, a hound moves from 8 to 6R forcing hare to retreat to 3R, after which the hound moves from 11 to 10C and regains control.

References and Further Reading

Martin Gardner, Mathematical Games: About two new and two old mathematical board games, *Sci. Amer.* **209** #4 (Oct. 1963) 124–130.

Martin Gardner, *Sixth Book of Mathematical Games from Scientific American*, W.H. Freeman, San Francisco, 1971, Ch. 5.

Édouard Lucas, *Récréations Mathématiques*, Blanchard, Paris, Vol. III, 1882, 1960, 105–116.

Sidney Sackson, *A Gamut of Games*, Random House, 1969.

Frederick Schuh, *Wonderlijke Problemen; Leerzam Tijdverdrijf Door Puzzle en Spel*, W.J. Thieme & Cie, Zutphen,1943,189–192.

Frederick Schuh, *The Master Book of Mathematical Recreations* (transl. F. Göbel, ed. T.H. O'Beirne), Dover Publications, New York, 1968, 239–244.

-22-

Lines and Squares

And I say to them, "Bears,
Just look how I'm walking in all of the squares!
And the little bears growl to each other, "He's mine,
As soon as he's silly and steps on a line."

A.A. Milne, *When We Were Very Young.*

On the square, to the left, was elegantly engraved in capital
letters this sentence: ALL THINGS MOVE TO THEIR END.

François Rabelais, *Pantagruel*, V, 37.

If you find you're bored to pieces with our other games, you should find your board and pieces to play these ones. The chapter contains several old friends and some new ones.

Tit-Tat-Toe, My First Go, Three Jolly Butcher Boys All in a Row

Oxford Book of Mother Goose Rhymes, 1951, p. 406.

The game is more usually known as Tic-Tac-Toe, or Noughts-and-Crosses, depending on which side of the Atlantic you are. Whoever moves first puts a cross (X) in one of the nine spaces in the board of Fig. 1. His opponent then puts a nought (O) into any other space and then they alternate X's and O's in the remaining empty spaces until one player wins by getting three of his own kind on one of the eight lines of Fig. 2. If teacher isn't listening he then shouts a

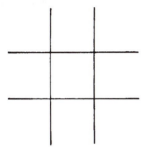

Figure 1. Tic-Tac-Toe Board.

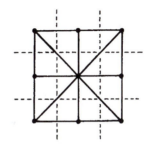

Figure 2. Its Eight Lines.

suitably triumphant phrase, which in some parts of America is

$$\text{"Tic-Tac-Toe, three in a row"},$$

and, in Holland, according to Fred. Schuh, is

$$\text{"Boter, melk en kaas, ik ben de baas"}.$$

When neither player is able to make a line we have a tied game. We have no doubt that most of our readers were bright enough as children to discover that this always happens when the game is properly played, and only the authors of books like *Winning Ways* retain sufficient interest to study the game in any detail.

But have you ever tried a complete analysis? If so, you've probably found that it took more space than you first thought it should. Later on we'll give a more concise analysis than most, though we admit that our rough work took more than one sheet of paper. But first let's look at three non-board games.

Magic Fifteen

In this game the players alternately select numbers from 1 to 9 and no digit may be used twice. You win by getting three numbers whose sum is 15. This game was suggested by E. Pericoloso Sporgersi.

Spit Not So, Fat Fop, as if in Pan!

Spit Not So, Fat Fop, as if in Pan! is a sentence for which we are indebted to Anne Duncan. It suggests the following game. Write the nine words on nine separate cards and have the two players alternately select cards, a player winning if he can collect all the cards that contain a given letter. This game was suggested by Leo Moser's game of **Hot** in which the nine words were HOT, FORM, WOES, TANK, HEAR, WASP, TIED, BRIM, SHIP, and the winner must collect *three* words with a common letter.

Jam

John A. Michon's game of **Jam** is played on Fig. 3. The players alternately select roads (straight lines) and whoever manages to take all the roads through a town wins.

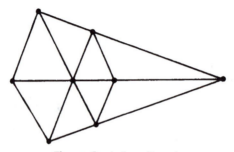

Figure 3. A Jam Board.

How Long Can You Fool Your Friends?

We'll bet you can fool most of them for quite a long time, playing any one of the above games. But they're all Tic-Tac-Toe in disguise, so you should be able to make the right moves while they are floundering! You can see why these games are all the same by arranging the numbers for Magic Fifteen as a magic square (Fig. 4(a)); the words for Spit, Etc. as in Fig. 4(b); and naming the towns or numbering the roads for Jam as in Fig. 4(c). For Hot you can prove the same thing by writing the words on Fig. 1 in the order we gave them. Can you find a better sentence than Anne's, possibly using redundant letters as in Hot? It would be nice if the words of your sentence could be written across the board in order! See the Extras.

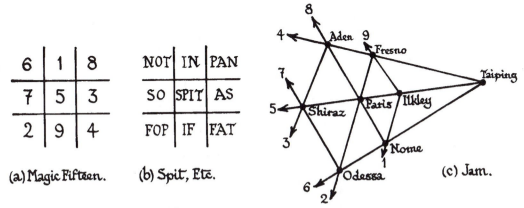

Figure 4. The Game's the Same by Any Name.

Analysis of Tic-Tac-Toe

For convenience we number the board as in Magic Fifteen and suppose by symmetry that the first move (X) is in 5 (Fig. 5), 6 (Fig. 6) or 7 (Fig. 7). We'll also suppose that each player is sensible enough to

(a) complete a line of his kind if he can, and

(b) prevent his opponent from doing so on his next move.

In the analysis,

bold numbers represent such **forced** moves,

! denotes a move that's better than some others,

? denotes a move that's worse than some others,

x denotes a win for Cross,

o denotes a win for Nought,

⊗ denotes a tied game,

~ denotes an arbitrary move, and

v. is a cross-reference to another column in the analysis.

The plays are given in numerical order, apart from the convention about the initial digit.

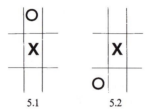

Figure 5. Starting in the Centre.

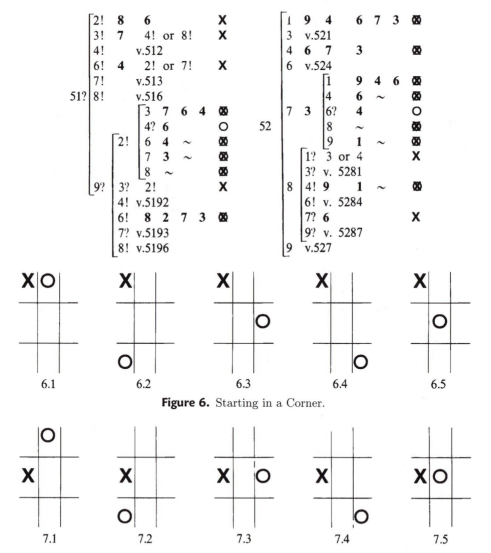

Figure 6. Starting in a Corner.

Figure 7. Starting on a Side.

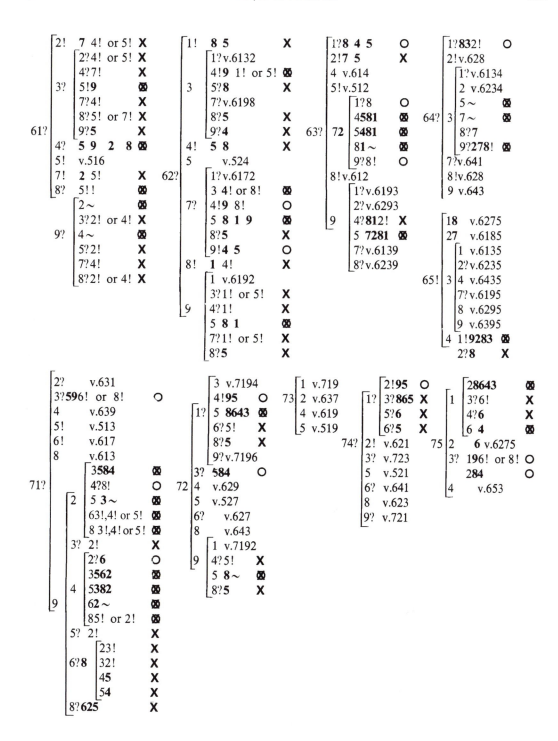

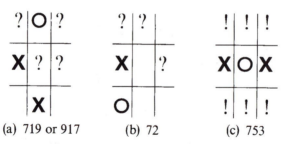

(a) 719 or 917   (b) 72   (c) 753

Figure 8. Lesser Known Byways of Tic-Tac-Toe.

In "The Scientific American Book of Mathematical Puzzles and Diversions" Martin Gardner remarks (and we agree) that many players have the mistaken impression that because they are unbeatable they have nothing more to learn. He gives three examples (Fig. 8) showing how a master player can take the best possible advantage of a bad play. In Fig. 8(a) X's last move was chosen so as to give O four losing chances out of six (the Enough Rope Principle). Against X's opening of 7, Gardner recommends O to reply with 2 since this offers X three losing chances (Fig. 8(b)). In Fig. 8(c) O can let X choose his move for him, since it is impossible for O to play without setting a winning trap!

Ovid's Game, Hopscotch, Les Pendus

In his *Ars Amatoria*, Ovid advises young women to learn certain games to amuse their lovers. He mentions in particular a certain *ludus terni lapilli* played on a *tabella* which is conjectured to be a moving form of tic-tac-toe, played with 3 black pebbles and 3 white ones. Several such games are known to have been popular in ancient China, Greece and Rome and in medieval England and France.

In the version nowadays known as **Ovid's Game**, the players take turns placing their pebbles on the board until all 6 are down. If neither player has won by getting the 3 of his kind in a row they continue playing by moving on each turn a single one of their pebbles to any orthogonally adjacent square. The first player has a sure win by playing in the centre:

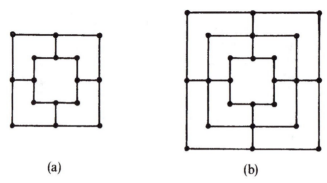

(a)   (b)

Figure 9. Six and Nine Men's Morris Boards.

$$5! \begin{bmatrix} 1 \ 4 \ 6 \ 8 \ 3, \ 4 \ \text{to} \ 9, \ \text{any}, \ 9 \ \text{to} \ 2, \ \text{or} \\ 2 \ 1 \ 9 \ 4 \ 6, \ 1 \ \text{to} \ 8, \ \text{any}, \ 5 \ \text{to} \ 3, \end{bmatrix}$$

so the central opening is usually forbidden, making the game a draw. However, the loopiness of the game allows many variations to occur in a single play and the game teems with traps.

We will use the name **Three Men's Morris** for the version in which, in the moving part of the game, the players are allowed any chess king move along the 8 lines of Fig. 2. An American Indian version, which has been called **Hopscotch**, allows any king move, whether on the 8 lines or not. It is a draw, even when the central opening move is allowed, as is the French version, **Les Pendus**, in which a pebble can be moved to *any* empty space.

Six Men's Morris

Six Men's Morris is played on the board of Fig. 9(a). Each player has 6 counters and the game has two phases as in Ovid's Game. First the counters are placed alternately by the two players. Then the counters are moved from one of the 16 nodes to an adjacent one along a line of the board. If a player gets three in a row he removes an opposing counter. A player wins when he reduces the opposing force to two counters.

Nine Men's Morris

Nine Men's Morris is played similarly with 9 counters for each player on a square or rectangular board designed as in Fig. 9(b). When a player forms a **mill** (gets three in a row) he again removes an opposing counter, but is not allowed to take one from an opposing mill. There are a number of variations and many names (Merrilees, Morelles, Mill, Mühle); see the books of R. C. Bell or H. J. R. Murray for details. Ralph Gasser has used an endgame database of about 10^{10} states and an 18-ply alpha-beta search to show that, with best play, Nine Men's Morris is a draw.

Three Up

This is a vertical three-in-a-row game. Each player starts with six checkers of his own color. They play alternately by putting a checker onto the table or onto a previous stack, and each tries to complete a stack three high of his own color. When all the checkers are placed, the players continue by alternately transferring single checkers of their own color from the top of one stack to the top of another, or possibly onto the table. At no time may any stack be more than three high.

It's very easy for a skilled player to beat a novice at this game, which has many cunning features. But Vasek Chvátal has shown that if you never try to win (by putting two of your pieces in a stack) then you can't lose! For if your opponent has set up $t \ (\geq 1)$ threats (stacks beginning with two of his pieces) then he can cover at most $6 - 2t$ of your checkers, so you have at least $2t$ uncovered ones—more than enough to deal with his threats.

Four-in-a-Row

It's clear that the first player can get 1-in-a-row on a 1×1 board, and 2-in-a-row on a 2×2 board, and we have seen that he can't get 3-in-a-row on a 3×3 board, but it's not hard to

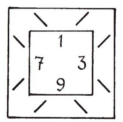

Figure 10. Four-in-a-Row is a Second Player Tie on a 5 × 5 Board.

show that he can get 3-in-a-row on any bigger board, even with just one extra square. How big a board is needed to get 4-in-a-row? C.Y. Lee observes that the second player can tie on a 5 × 5 board. His strategy is to play as in Tic-Tac-Toe whenever the first player plays in the central 3 × 3 square. You won't have too much difficulty if you remember this and note that when you play in the squares marked with a diagonal line in Fig. 10 you sabotage your opponent's chance of getting 4-in-a-row on the border, and also on a diagonal involving two of the squares 1, 3, 9 and 7.

Lustenberger has used a computer to show that 4-in-a-row is a win for the first player on a 4 × 30 board.

By far the most interesting and popular version is the 3-dimensional one, played on a 4 × 4 × 4 cube. Oren Patashnik has shown that the first player can always win at 4 × 4 × 4 **tic-toc-tac-toe**. Patashnik's solution now includes a computerized dictionary of several thousand openings. This dictionary was obtained by patient and skilful interaction between Patashnik and a computer over a period of many months. It is too large to be accessible other than by computer. Several skeptical computer scientists have recently examined Patashnik's dictionary and it is now accepted as complete and correct.

Five-in-a-Row

It's quite a good game just to try to get 5 in a row orthogonally or diagonally on any reasonably large board. Mathematicians will prefer to play **Five-in-a-Row** on an infinite board.

In this kind of game there are several well defined degrees of threat and when playing with children and good friends it's nice to announce these by suitable cries. We recommend

SHOT! for a threat to win next move, e.g.

SHOTS! for two or more SHOTs at once, e.g.

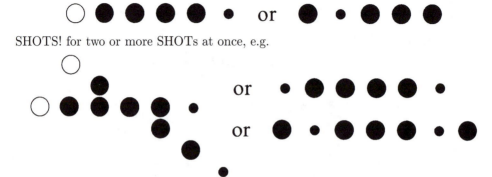

POT! if you can guarantee a SHOT next move, e.g.

POTSHOT! for a POT and a SHOT at the same time, e.g.

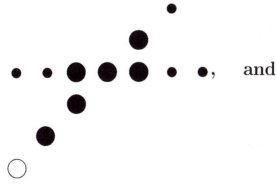

POTS! for two or more POTs at once, e.g.

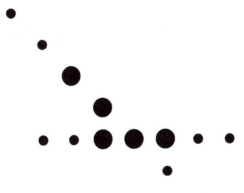

These can be of great help in understanding the effects of forced moves, for example:

A SHOT, typically a line of 4 open at one end, must be blocked instantly. So a pair of SHOTs wins next move.

A POT, typically a line of 3 open at both ends, must either be blocked immediately or staved off with a SHOT. So a POTSHOT wins unless possibly the move that blocks the SHOT is a countering SHOT. Against a pair of POTs you can only hope to defend by making a sequence of SHOTs until one of them happens to block one of the POTs.

These terms can be applied to many games of this type, for instance in 4-in-a-row,

is a SHOT and

is a POT. Similar cries are used in Phutball (see later in this chapter); there are obvious connexions with the notion of *remoteness* in Chapter 9.

Five-in-a-row has been called Go-Bang in England for at least a hundred years and has more recently been called Pegotty or Pegity (Parker Bros., U.S.A.).

Go-Moku

In Japan there are several perfect players who can always claim their win in their version, **Go-Moku**, of 5-in-a-row on a Go board, size 19×19, even though the first player is handicapped by not being allowed to make the **fork threat** of a pair of open lines of 3 (we'd cry POTS! for this) and *six* in a row is *not* counted as a win. Allis, van der Herik and Huntjens have used a computer to show that Go-Moku is a first player win.

Six, Seven, Eight, Nine, . . . , in a Row

A. W. Hales and R. I. Jewett have produced an ingenious pairing strategy which shows that many games of this type are tied or drawn. For instance here is a quick proof that 5-in-a-row is tied on a 5×5 board. All you have to do is to make sure that for every move of your opponent in a marked square in Fig. 11 you take the similarly marked square in the direction indicated by the mark. So you could give her the centre square *and* let her make the first move as well. If the position you're presented with already satisfies the condition, make a random move. At the end of the game there will be at least one of your counters in every conceivable winning line.

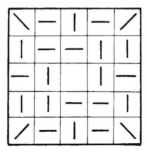

Figure 11. A Hales-Jewett Pairing.

You can see that 9-in-a-row is a draw on an infinite board with the Hales-Jewett pairing of Fig. 12. When your opponent takes the cell at one end of a line in the figure, you take the one at the other. The result was first proved in 1954 by Henry Oliver Pollak and Claude Elwood Shannon, using the following strategy. Tile the board with H-shaped heptominoes: the second player plays ordinary tic-tac-toe in each of these regions, concentrating on preventing a line of 3 in either a diagonal, or the horizontal, or the right vertical. John Lewis Selfridge also gave a Hales-Jewett pairing on an 8×8 board, which could be used to tile an infinite one and give the same result.

T. G. L. Zetters (nom de guerre of some Amsterdam combinatorists) recently showed that the second player can even draw 8-in-a-row. Their proof uses a parallelogram-shaped tile of 12 cells, and goes some way towards showing that 7-in-a-row is also a draw.

Figure 12. Nine-in-a-Row is a Draw on an Infinite Board.

S.W. Golomb has found a Hales-Jewett pairing for 8-in-a-row on an $8 \times 8 \times 8$ cube. It will be easier to explain if we first describe the analogous two-dimensional solution for 6-in-a-row on a 6×6 square. Figure 13(a) is like Fig. 11 except that you may reply to a move on a diagonal with *any* other move on the same diagonal. Note that the figure has the mirror symmetries indicated by the two thick lines, so it would suffice to indicate only one quadrant as in Fig. 13(b).

Figure 14 indicates one octant of Golomb's $8 \times 8 \times 8$ pairing in a similar way. The symbols $-$, $|$, and $\backslash$ are in the horizontal layer shown, while $\bullet$ is all you can see of a vertical line. The arrows pierce the layers and represent lines of obvious directions in various diagonal planes. The three mid-planes of the $8 \times 8 \times 8$ cube (represented by the thick lines in Fig. 14) are reflecting planes and, as in the 6×6 pairing of Fig. 13, you may respond to *any* move on a body diagonal with another on the same diagonal. In fact Golomb can give you *any* six cells on *each* of the four body diagonals *and* allow you to have first move and still tie the game.

Golomb and Hales have obtained further results on hypercube Tic-Tac-Toe.

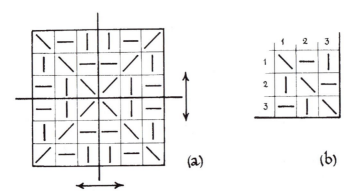

Figure 13. A Pairing for Six-in-a-Row on a 6×6 Board.

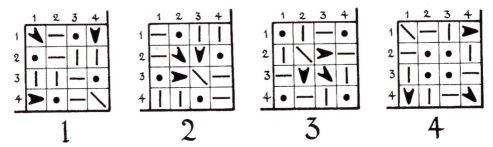

Figure 14. Golomb's Pairing for Eight-in-a-Row on an $8 \times 8 \times 8$ Cube.

n-Dimensional k-in-a-Row

Hales and Jewett consider the game of k-in-a-row on the n-dimensional

$$k \times k \times k \times \ldots \times k$$

board. They prove that if k is sufficiently large, namely

$$k \geq 3^n - 1 \qquad (k \text{ odd}) \text{ or}$$
$$k \geq 2^{n+1} - 2 \qquad (k \text{ even})$$

the game is tied by a suitable pairing strategy, and on the other hand that if n is sufficiently large compared to k, it is a first player win by the strategy stealing argument described below. They conjecture that the game is tied if there are at least twice as many cells as lines.

How many lines are there? Leo Moser has remarked that each line is determined by either one of the two cells which extend it into the surrounding

$$(k + 2) \times (k + 2) \times (k + 2) \times \ldots \times (k + 2)$$

cube, so that the total number of lines is exactly

$$\frac{1}{2}\{(k+2)^n - k^n\}.$$

The Hales-Jewett conjecture is therefore that the game is tied whenever

$$k^n \geq (k+2)^n - k^n,$$

i.e.

$$2k^n \geq (k+2)^n.$$

So it should be true if $k \geq 3n$, for example; Leo Moser has proved that it's true if $k > cn \log n$ for some constant c.

Strategy Stealing in Tic-Tac-Toe Games

For almost all forms of tic-tac-toe game there is a strategy stealing argument which shows that the second player cannot have a winning strategy. Though earlier authors probably knew it, this was formally proved by Hales and Jewett. We suppose that each player has an indefinite supply of his own kind of piece, that the pieces don't move after they're once put down, and that each player's aim is to produce a winning configuration with some of his pieces.

The assertion is that all such games in which the winning configurations for the two players are similar, are either wins for the first player or are tied under best play. For if the second player had a winning strategy, then the first player could steal it as follows. After a random first move he could pretend to *be* the second player, ignoring his opening move and making a random move whenever the stolen strategy would otherwise repeat a move already made. We conclude that if the second player had a winning strategy, so would the first, since an additional piece on the board can never harm him! Obviously *both* players can't win at once, so the supposed winning strategy for the second player cannot exist.

The argument applies to n-in-a-row on any shape of board, provided no special restrictions, like those of Go-Moku, are added. In this case the winning configurations are just the appropriate lines of n, and are exactly the same for each player.

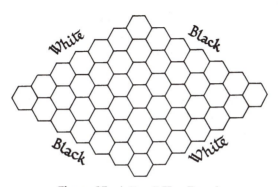

Figure 15. A 7×7 Hex Board.

However, in the most notorious cases of strategy theft the winning configurations are not identical but related by a symmetry of the board (and so are still *similar*). These are listed below, in chronological order.

Hex

Hex is played on a rhombus of hexagons like that of Fig. 15. Black wins if his pieces connect one pair of opposite sides of the board and White if his connect the other pair.

Hex was invented by Piet Hein and the strategy stealing argument found by John Nash. Cameron Browne has written a very good book on Hex strategy. Vadim Anshelevich has written a tournament-winning Hex-playing program.

Bridgit

Bridgit (or Gale) is played on two interlaced n by $n + 1$ lattices. Left joins two adjacent (horizontal or vertical) spots of the black lattice and Right makes similar moves in the white one. No two moves may cross. In Fig. 16 Left has just won since he has formed a chain connecting a topmost spot to a bottommost one. Bridgit was invented by David Gale and its strategy stealing argument by Tarjan.

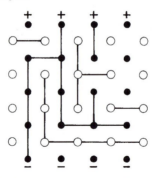

Figure 16. Left Forms a Black Chain in Bridgit.

How Does the First Player Win?

In these cases, as in the examples considered by Hales and Jewett, it is impossible for a completed game to be tied, so that the argument actually proves that the first player can win, but does not give much help in finding an explicit winning strategy for him. No explicit strategy for Hex is known, and Tarjan and Even have shown that, in the technical sense, generalized Hex is hard. But for Bridgit an explicit pairing strategy was found by Oliver Gross, and many other strategies can be deduced from Alfred Lehman's subsequent theory of the Shannon Switching Game.

The Shannon Switching Game

The Shannon Switching Game generalizes Bridgit. It is played on a graph representing an electrical network in which certain nodes are labelled + and some others are labelled −.

Each edge (begin the game with them drawn in *pencil*) represents a permissible connexion between the nodes at its ends. *Mr. Shortt*, at his move may *establish* one of these connexions permanently (*ink over* a pencilled edge) and attempts to form a chain between some + node and a − one. His opponent, *Mr. Cutt* may permanently *prevent* a possible connexion (*erase* a pencilled edge) and tries to separate + from − forever. Figure 17(a) shows a Shannon game equivalent to our Bridgit one. You can always suppose that there's only one positive node and one negative one by making identifications as in Fig. 17(b).

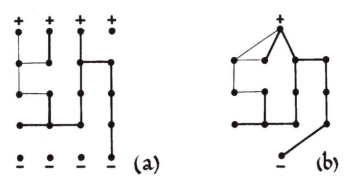

Figure 17. Bridgit Played as a Shannon Switching Game.

Supposing this, Lehman has proved that Mr. Shortt can win as second player if and only if he can find two edge-disjoint trees that each contain all the nodes of some subgraph containing + and − . The "only if" part is hard, but there's an easy strategy which proves "if": whenever Mr. Cutt's move separates one of the trees into two parts, A and B, Mr. Shortt makes a move on the other tree joining a vertex of A to one of B.

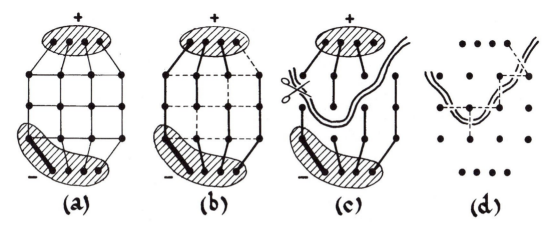

Figure 18. How Mr. Shortt Wins a Game of Bridgit.

Let's use Lehman's theory to show how the first player, who should regard himself as Mr. Shortt, can win in Bridgit. After his first move (Fig. 18(a)) Mr. Shortt (who's now *second* in line to move) can see the two edge-disjoint trees indicated by the thick and pecked lines of Fig. 18(b) (remember to regard each of the + and − sets as connected, including the node that has been Shortted to −). If now Mr. Cutt disconnects one of the trees, for example by erasing the scissored edge in Fig. 18(c), then Mr. Shortt should secure one of the six bridges (Fig. 18(d)) across the imaginary river that now separates the two parts of the tree severed by Mr. Cutt.

The game can be generalized to make the winning configurations for Mr. Shortt just those which contain a specified family P of sets of edges. (In the original game P was the family of paths from + to − .) Lehman proves the "only if" part of his theorem by taking P to be the family of all trees containing every vertex (spanning trees).

If Mr. Shortt, as second player, has a win in the modified game, it's *very* easy to see that there must be two edge-disjoint spanning trees. For since an extra move is no disadvantage, both players can play Mr. Shortt's strategy! If they do this, *two* spanning trees will be established, using disjoint sets of edges. Conversely, if two such trees exist, our previous strategy for Mr. Shortt actually wins for him as second player, even in the modified game.

The more detailed part of Lehman's argument establishes that, in a suitable sense, the modified game reduces to the original one.

The Black Path Game

This elegant little game was invented by Larry Black in 1960. You can play it on a rectangular piece of paper ruled into squares as in Fig. 19. At any time the squares that have been used will each contain one of the three patterns shown in Fig. 19(b) and will include a path like the black path in Fig. 19(a) which begins at the starting arrow. The player to move must continue the black path by drawing one of the permissible patterns in the next square. You lose if your move makes the black path run into the edge of the board. The numbers 1 to 8 show the order of the first eight moves in our sample game, and the next player must now move in the square marked 9. You'll see that pattern 1. loses instantly, 2. wins quickly and 3. loses slowly.

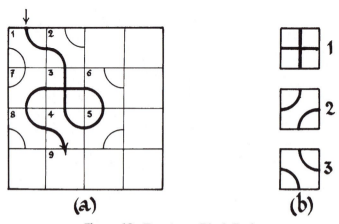

(a)   **(b)**

Figure 19. Forming a Black Path.

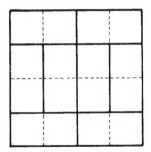

Figure 20. Black Path Game Board Divided into Dominoes.

We have a pairing strategy by which the first player can win on any rectangular board with an even number of squares. He imagines the board divided into 2×1 dominoes in any way he likes, for instance Fig. 20, and then plays so as to leave the end of the path in the middle of a domino (which can never be the edge of the board!). On an odd by odd board it is the second player who can win, by dividing all of the board except the opening square into dominoes.

Lewthwaite's Game

Domino pairing (e.g. Fig. 21(b)) also enables the second player to win a game invented by G. W. Lewthwaite in which 12 white and 12 black squares are slid alternately in a 5×5 box from the starting position of Fig. 21 (a), and a player who, at his turn, cannot move any piece of his color, loses. What happens if a player is also allowed to slide a row or column of 2, 3 or 4 squares, provided both end squares are of his color?

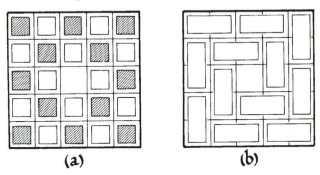

Figure 21. Pairing Gives a Second Player Win in Lewthwaite's Game.

Meander

Meander was also invented by Lewthwaite and is also played with 24 tiles in a 5×5 box, but the tiles are now patterned as in Fig. 22(a), and are slid by either player. Figure 22(b) shows the starting position. The winner is the first player to produce a continuous curve connecting the boundary to itself and involving at least three tiles, as in Fig. 22(c). There are two versions of the game. In the first, players alternately slide just one tile; in the other, a row or column of 1, 2, 3 or 4 tiles may be slid as a single move.

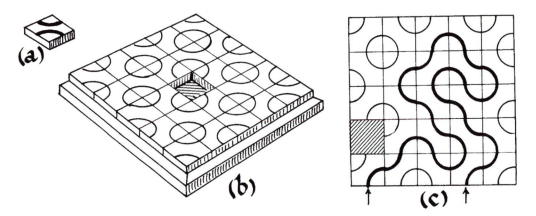

Figure 22. Meander.

Winners and Losers

Frank Harary has proposed a family of games, one for each polyomino, P, all played on an infinite board. On alternate turns, Left makes a square black, and Right makes one white, and Left's aim is to produce a black copy of P, while Right tries to foil him. Harary calls P a **winner** if Left has a winning strategy—otherwise a **loser**.

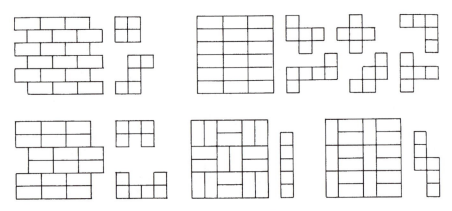

Figure 23. Hales–Jewett Pairings Make Twelve Polyomino Losers.

The twelve polyominoes of Fig. 23 can be proved to be losers using the indicated Hales-Jewett pairings, mostly found by Andreas Blass. If P contains one of these it is therefore a loser. Martin Kutz notes that the lower left pairing does not suffice for the U-Hexomino, and that the appropriate pairing consists of staggered sets of three rows in place of the sets of two. The only polyominoes *not* containing one of these twelve are the twelve shown in Fig. 24. Eleven of these are known to be winners, with known strategies which win in m moves on a $b \times b$ board (see the figure). The last, called "snaky" by Harary, is also conjectured to be a winner.

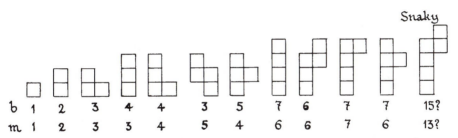

Figure 24. Twelve Polyomino Winners with Board Sizes and Numbers of Moves (but Snaky's a bit shaky).

The game of Pentominoes, played with the 12 pentominoes on an 8×8 board (if you can't place a non-overlapping piece, you lose) has been shown to be a first-player win by Hilarie Orman, who suggests that a computationally challenging problem would be to solve the similar game with the 35 hexominoes on a 15×15 board.

Dodgem

Colin Vout invented this excellent little game played with two black cars and two white ones on a 3×3 board, starting as in Fig. 25(a). The players alternately move one of their cars one square in one of the three permitted directions (E, N or S for Black; N, E or W for White) and the first player to get *both* of his cars off the board wins. Black's cars may only leave the board across its right-hand edge and White's cars only leave across the top edge. Only one car is permitted on a square, and you lose if you prevent your opponent from moving.

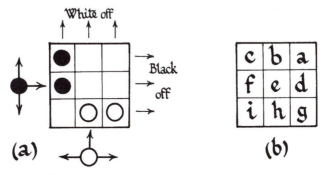

Figure 25. Colin Vout's Game of Dodgem.

Although the board is the same size as that for Tic-Tac-Toe, this game is much more interesting to play. Table 1 contains the outcome of every position; the column gives the positions of the black cars, the row those of the white, labelled by pairs of letters from Fig. 25(b). A blank entry represents an illegal position, since only one car is allowed on each square.

+ is a win for Black (Left),
− is a win for Right (White),
O is a win for the second player,
∗ is a win for the first player.

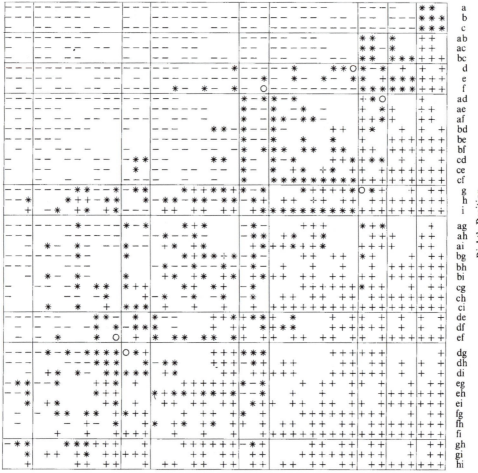

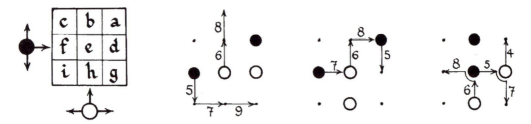

Table 1. Outcomes of Positions in Dodgem and Some Good Moves.

If you haven't got our table drawn on the back of your hand, you'll find this little game hard to play against an expert, who'll spring all sorts of little traps for you. It's often *not* a good idea to push your car off as soon as you can since it may be more useful blocking your opponent. In many situations it's a good idea to aim for the top right hand corner.

When you are expert, you can try playing Dodgem with $n-1$ cars of each color, on an $n \times n$ board, starting in the first column and row, with the SW corner empty.

Dodgerydoo

This game is played with two Dodgem cars on a quarter-infinite board. Now either player may move either car any distance North or West in a single move, provided it does not jump on to or over the other car. If you can't move you lose.

It's not hard to see that

> any position in which
> the two cars are
> on neighboring squares
> is a $\mathcal{P}$-position,

because whatever the next player does you can continue to shadow him. As a consequence,

> any other position with
> the cars in the
> same row or column,
> or in adjacent ones,
> is an $\mathcal{N}$-position,

because the next player can immediately creep one car up to the other. So in analyzing later positions we might as well make it illegal to have both cars in the same row or column, or in adjacent ones.

Let (x_1, y_1) and (x_2, y_2) be the positions of the two cars in this restricted game. Then on these numbers we are playing a nim-like game with four heaps in which we can reduce any one of the four numbers x_1, x_2, y_1, y_2 provided we ensure that neither $x_1 - x_2$ nor $y_1 - y_2$ is 0 or ± 1. Since the x's and the y's don't interact, we can regard this as the sum of two games, one played on the x's, the other on the y's. Table 2 gives the nim-values for either of these games—apart from the positions described in the boxes above. An X denotes an illegal position in the restricted game.

> The position (x_1, y_1), (x_2, y_2)
> is a Dodgerydoo $\mathcal{P}$-position
> just if $f(x_1, x_2) = f(y_1, y_2)$,

(since their nim-sum is then 0), where $f(x_1, x_2)$ is the function given in Table 2.

x_1 \ x_2	0	1	2	3	4	5	6	7	8	9	10	11	12	13	14	15	16
0	X	X	0	1	2	3	4	5	6	7	8	9	10	11	12	13	14
1	X	X	X	0	1	2	3	4	5	6	7	8	9	10	11	12	13
2	0	X	X	X	3	1	2	6	4	5	9	7	8	12	10	11	15
3	1	0	X	X	X	4	5	2	3	8	6	10	7	9	13	14	11
4	2	1	3	X	X	X	0	7	8	4	5	6	11	13	9	10	12
5	3	2	1	4	X	X	X	0	7	9	10	5	6	8	14	15	16
6	4	3	2	5	0	X	X	X	1	10	11	12	13	6	7	8	9
7	5	4	6	2	7	0	X	X	X	1	3	11	12	14	8	9	10
8	6	5	4	3	8	7	1	X	X	X	0	2	14	15	16	17	18
9	7	6	5	8	4	9	10	1	X	X	X	0	2	3	15	16	17
10	8	7	9	6	5	10	11	3	0	X	X	X	1	2	4	18	19
11	9	8	7	10	6	5	12	11	2	0	X	X	X	1	3	4	20
12	10	9	8	7	11	6	13	12	14	2	1	X	X	X	0	3	4
13	11	10	12	9	13	8	6	14	15	3	2	1	X	X	X	0	5

Table 2. Dodgerydoo Values, $f(x_1, x_2)$

There doesn't seem to be much pattern in the table, once we get away from the edge, but at least the first few rows (and columns) are arithmetico-periodic. The (ultimate) periods and saltuses in the first five rows are 1, 1, 3, 9, 36, and are valid outside the heavy line. The same table solves two-car Dodgerydoo in three dimensions, for which the $\mathcal{P}$-position condition becomes

$$f(x_1, x_2) \overset{*}{+} f(y_1, y_2) \overset{*}{+} f(z_1, z_2) = 0.$$

Philosopher's Football

Philosopher's Football or PHUTBALL (registered J.H. Conway) for short, is a very playable game that you can read about for the first time in this book. It is usually played on the 15×19 intersections of the board shown in Fig. 26, or on a 19×19 Go board, using one black stone (the **ball**) and a large supply of white ones (**men**). All pieces are common to both players and indeed both players have the same legal moves although their aims are different.

Start with the pitch empty except for the ball which starts at the central spot. Then each player, when he moves, must

　　　　　either place a new man at any unoccupied intersection
　　　　or　　**jump** the ball, removing the men jumped over.

(He may not do both.)

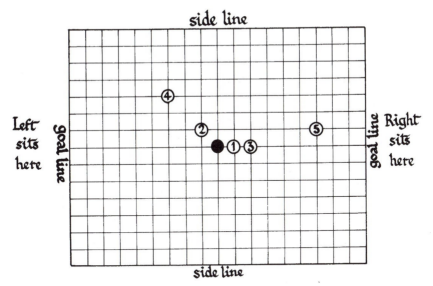

Figure 26. The Phutball Pitch and the First Five Moves of a Game.

A single jump of the ball may be in any of the eight standard compass directions, N, NE, E, SE, S, SW, W, NW on to the first empty point in that direction provided at least one man is jumped over. All the men jumped over are removed instantly. A player may take several consecutive such jumps in various of the eight directions as a single move. But because the men are removed instantly, the same man cannot be jumped over more than once in a move, and no man can be placed on the board in a jumping move.

It is legal for the ball to land on any of the goal lines or side lines. It is also legal for the ball to leave the board, but only by jumping over a man on the goal line, and only as the last move of the game. In fact Left's aim is to arrange that at the *end* of a move the ball is either *on or over* Right's goal line, while Right's is to get it on or over Left's. However a defender can sometimes successfully use his own goal line by jumping the ball onto and off it during a single move. In the standard opening,

Left, Right, Left, Right, Left,

will place the stones

1,   2,   3,   4,   5,

of Fig. 26, building chains towards their opponents' goals. Right is now frightened by Left's threat to make a long jump over 1 and 3 and later establish a chain through 5. He therefore makes two short jumps himself over 2 and 4.

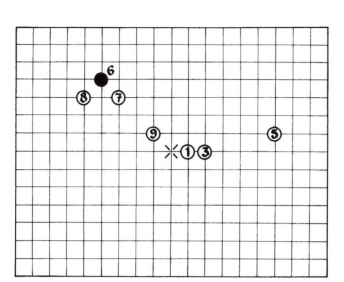

Figure 27. The Next Four Moves.

In the subsequent moves

$$7, \quad 8, \quad 9,$$

of

$$\text{Left, Right, Left,}$$

Left tries to reestablish his chain while Right prepares the way for a sideways jump to defend against this. If it were Left's turn to move in Fig. 27 he could in one move make two jumps over 7 and 9 and a longer jump over 1 and 3 (it would probably be better for him not to make this last jump; as in Chess, a threat is often more powerful than its execution). However, it's Right's turn so he jumps over 8, and the next few moves are shown in Fig. 28.

These are all rather subtle. Left's move 11 is much better than reinstating 8, which Right could too easily **tackle** by placing a man where the ball was in Fig. 27 (after a jump of these two stones, Left would find it very difficult to reestablish a useful connexion with the rest of his chain). Right's move 12 is even more subtle! A direct threat to win at this point would make Left jump over 11 and 7, and arrive at a commanding position. Move 12 provides a way back after this jump and also prepares the way for a move at 14, followed by a roundabout triple jump over 11, 12 and 14, which both gets Right near to the Left goal line and removes some pieces useful to his opponent. The move 12 has even more hidden secrets: if Left places 13, *Right* can make the jump over 11 and 7, and then any Left threat to connect with his old chain equally helps Right to connect with 13 and 12.

Almost all these moves have become standard, but from now on experts differ. The game has many subtle tactics (tackling, poisoning one's opponent's threats, devastating U-turns, ...) and we'll only offer a few hints.

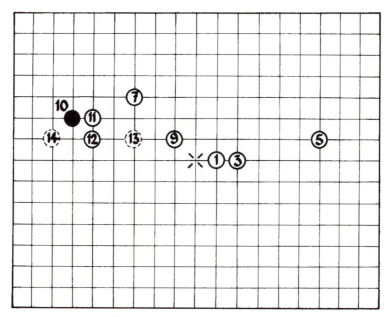

Figure 28. The Game Continues.

Try not to jump until you really have to, and then only as far as you really must. If you will have a stone within three of the place your opponent will jump to, but *not* a knight's move away, you can probably use it to get back and needn't be too frightened by his jump (which he probably shouldn't be making!). Remember that a stone a knight's move away from the ball is almost always useless. Such stones are called **poultry** (a corruption of paltry and parity). A threatened chain becomes much more useful if it can be jumped along in several different ways. Don't forget that the stone you place may be useful to your opponent—possibly in a devastating U-turn.

A pleasing feature of the game is that an expert can still enjoy a game against a novice provided they start with the ball much nearer the expert's goal-line.

Like Chess and Go, and unlike most of the games in this book, Phutball is *not* the kind of game for which one can expect a complete analysis. In fact, Demaine, Demaine, and Eppstein have shown that the problem of determining whether a player has a win on the next Phutball move is NP Hard. Even one-dimensional Phutball has not been completely analysed, although a restricted version of the one-dimensional case has been solved by Grossman and Nowakowski.

Extras

Count foxy words And stay awake Using lively wit

is a correspondent's ingenious answer to Anne Duncan's question.

Amazons

Amazons was invented by the Argentine Walter Zamkauskas in 1988, and first published in the magazine *El Acertijo* in 1992. In 1993, Michael Keller, the editor of *World Games Review*, introduced it to a postal gaming club called the Knights of the Square Table (NOST). In January 1994, an English version of the rules (translated by Keller) appeared in *World Games Review* #12. The first international tournament was played by fax in 1994–95 between Argentina and the U.S., which ended in a 3-3 tie. The game is played on the squares of a square board, usually 10×10 or 8×8.

An Amazon is an immortal chess queen. Each side begins with several, usually 4. At each turn, a player selects one of his Amazons, moves her, and then completes the turn by shooting an arrow from the Amazon just moved. This arrow also moves like a chess queen, as far as desired in a straight line, either horizontal, vertical, or diagonal. The arrow "burns out" the square on which it lands, often denoted by placing a black Go stone there. Burned out squares are unplayable for the rest of the game. Neither Amazons nor arrows may jump over other Amazons or over burned out squares. The game ends when one player is unable to move because all possible moves of all of her Amazons are blocked.

Latestage Amazon endgames often decompose into sums of disjoint active regions. Each such region has a conventional value. Berlekamp [2000] and Snatzke [2002] have shown that the canonical forms are often very complicated, but that their thermographs are often quite tractable.

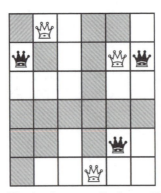

Figure 29. An Amazons Endgame Problem.

Although many Amazons tournaments have been played on the Internet, no significant literature has yet appeared. Most players think the first player enjoys some advantage, although there is a wide range of opinion about how big that advantage might be.

Who can win the Amazons endgame shown in Fig. 29? Black, White, First, or Second? Answer can be found on page 761 at the end of this section.

Checkers

Checkers is a very popular classic board game, also known as Draughts. It has long attracted the interest of artificial intelligence experts, most notably Jonathan Schaeffer [1992]. Although positions do not tend to decompose, Berlekamp [2002] has composed a problem, a sum of positions in Checkers, Chess, Go, and Domineering, that emphasizes subtle differences between the values of different winning checkers positions.

Chess

Chess has perhaps attracted more attention from artificial intelligence experts than any other game. Although most positions do not decompose into smaller pieces that can be analysed in the style of *Winning Ways*, Noam Elkies [1996, 2002] has discovered and composed many fascinating ones that do.

Cherries

The vertices of a graph are coloured blue, red or green. Right removes a vertex of minimal degree which is either red or green and Left removes a vertex of minimal degree which is either blue or green. Note: if all the minimal degree vertices are blue then Red does not have a move. Normal Play rules. Large classes of positions have been studied with values known to be integers, half-integers, and star.

Clobber

Clobber was invented in the summer of 2002 by Michael Albert, J. P. Grossman, and Richard Nowakowski. The first tournament was held at an academic conference on combinatorial game theory in Daghstuhl, Germany, in February 2002.

The legal moves of any piece are the intersection of a chess king and a chess rook, i.e., one square in any of four directions. We call such a piece a "duke." The game begins with all squares on a rectangular board occupied according to their colors on a checkerboard: each white square occupied by a white duke, and each black square by a black duke. Capturing occurs as in chess, one moves onto the square occupied by the opponent's duke, which is then removed from the board. But in Clobber, every move MUST be a capture. It is easily seen that the value of the game is all-small, because either side has a legal move just if there is some pair of adjacent pieces of opposing colors.

Clobber endgame positions usually decompose into sums. A wide and fascinating range of atomic weights has been observed.

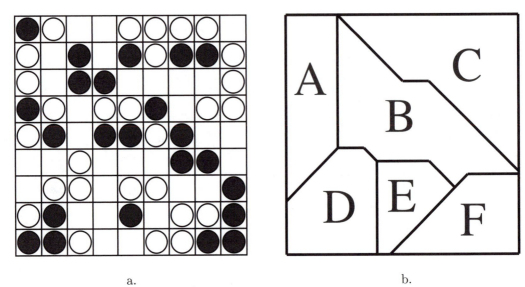

a.   b.

Figure 30. A Clobber Endgame Problem.

The first international Clobber problem composition contest concluded in August 2002. The winning entry, composed by Adam Duffy and Garrett Kolpin, is shown in Fig. 30. Who can win? Black, White, First, or Second? Answer can be found on page 761 at the end of this section.

Go

Go is THE classical Asian board game, which has been popular for many hundreds of years in Japan and for several millenia in China. The many active Go clubs and sponsored tournaments throughout Korea, Japan, China, and Taiwan support well over a thousand active professionals. There are now at least 50 active computer Go-playing programs, which compete against each other in major annual tournaments. Go has proved even more challenging to artificial intelligence researchers than chess; every program yet written can now be routinely defeated by its author and perhaps about one million other human Go players who are ranked 5 kyu or better.

Although there are more than a half-dozen modern dialects of the rules, they all agree so often that endgame positions whose outcomes depend on the fine points of the rules are very rare. Loopy positions (called Kos and superkos) are possible, but the commonly occurring such possibilities are all prohibited by every dialect of the rules. Virtually all latestage endgame positions decompose into sums. Values tend to be relatively complicated, but they can often by simplified by chilling (as defined on page 187 in the Extras to Chapter 6), a transformation that is nearly always reversible because Go positions satisfy special Diophontine constraints, i.e., their stopping positions must not only be dyadic rationals; they must be integers.

Generalized thermography provides a very powerful tool for evaluating earlier endgame positions. The folklore that professional Go players have taught for generations includes methods which we now view as effective techniques for finding approximate thermographs of Go positions quickly. Current research in Mathematical Go addresses earlier endgame positions, many of which can often be evaluated by modifications of the decomposition methodology even when they do not strictly decompose.

Professional experts and tournament directors are now approaching a global consensus that the value of playing first is worth about 6 or 7 points. Although most of the Go teachings are orally transmitted from each generation of professional players to their pupils, there are over a thousand books in Asian languages, over a hundred of which have been translated into English, and one book in English that has been translated to Japanese.

Konane

Konane, also called "Hawaiian Checkers", is the classic board game of ancient Hawaii. Like Clobber, the game begins with all squares occupied: white stones on all the white squares and black stones on all the black squares. Then one pair of adjacent stones is removed from the centre of the board or as near to the centre as possible. Thereafter, the legal move is to select one of your stones and jump (either horizontally or vertically) over an opposing stone. As in checkers, the jumped stone is removed from the board. Multiple jumps in a straight line are allowed, but multiple jumps that would turn at right angles are prohibited. The game ends when one player is unable to move.

Several state parks in Hawaii have displays showing relics of ancient boards and sets of stones gathered from nearby beaches: white stones were coral and black stones were volcanic.

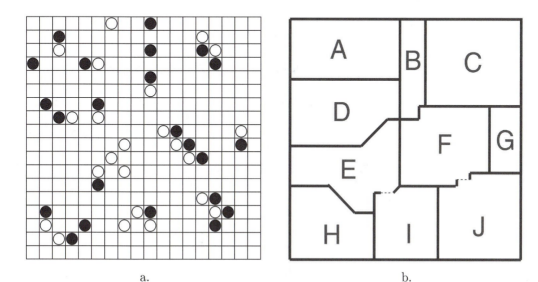

a.   b.

Figure 31. A Konane Endgame Problem.

A popular board size was 18×18. But unfortunately, the ancient Hawaiians had no writing. Although there are reports of historical games played in the king's palace, no written records of any expert games survive. The game has declined in popularity in modern Hawaii, and there does not seem to be any surviving folklore about what might constitute a strong opening strategy in this game, nor about how large an advantage, if any, is enjoyed by the player who moves first.

Latestage endgame positions decompose into sums. Many familar values occur. The reader may enjoy working out the Konane endgame problem shown in Fig. 31. Who can win? Black, White, First, or Second? Answer can be found on page 762 at the end of this section.

Reversi-Othello

The game of Reversi was invented by Lewis Waterman in England in 1888, and has been more recently popularized as Othello. It is usually played on the squares of an 8×8 board. Pieces are disks, white on one side and black on the other. In the initial position, there are 4 pieces in the centre of the board, 2 white and 2 black. A legal move consists in placing a piece of your color onto an empty square of the board and capturing a vertical, horizontal or diagonal sequence of contiguous pieces enclosed between opposing pieces. The captured pieces are turned over to become the color of the captor. Moves that fail to capture anything are illegal. The game ends when a player is unable to move because all pieces on the board are of the opponent's color, or because the board is fully occupied. In the latter case, the score is the number of pieces of each color in the final position.

Othello positions do not decompose into sums. The game has attracted considerable attention from the artificial intelligence and complexity theory communities, who have proved it to be PSPACE complete via transformation from Generalized Geography played on bipartite graphs with maximum valence 3.

Scrabble

Scrabble is a popular word game in which players draw letters printed on tiles and try to score points by playing them onto a board in patterns that form words in the dictionary. Because players cannot see either their opponents' tiles nor those not yet drawn, the game lacks complete information. For this reason, as well as rules which are defined only via a lengthy dictionary, this game would appear not to qualify for inclusion in *Winning Ways*. However, there is now a new genre of Scrabble problems that we feel merit the attention of our readers. These are positions in which you must find a way to win the game *no matter* which unseen tiles are held by your opponent and which remain untaken. Although most popular scrabble players try only to think of the play that will maximize their additional score on the next turn, these problems require strategies which plan several moves ahead.

Shogi

Shogi is Japanese chess. The board is slightly larger and the pieces have somewhat different moves than in Western chess. But the most significant difference is the possibility of "dropping" a previously captured Shogi piece back onto the board under a much wider variety of conditions than pawn promotions in western chess.

Sowing Games

There are many such games, notably Mancala, Wari (variously spelt), Ayo and Tchouka(illon). Bulgarian Solitaire, popularized by Martin Gardner and solved by Igusa, also comes under this head, as do so-called chip-firing games that have attracted the attention of combinatorialists and complexity theorists, and there is now a considerable literature. In a typical sowing game there is a sequence or cycle of bowls, some of which may contain seeds. A move is to take some seeds and 'sow' them, one or more at a time, into the bowls. Often the object is to accumulate the majority of seeds in specified bowls.

Answers to Problems

The Amazons position in Fig. 29 is the sum of three independent regions: the Northwest (NW), the northeast (NE), and the South (S). Treating bLack as Left, the values of these regions are *, $+1/4$, and $-1/2$ respectively. So White can win. If Black goes first, a dominant line of play is Black in the NW, then White in NE, Black in NE, and White in S, after which the position is as shown in Fig. 32 and each regional value is 0.

The Clobber endgame problem of Fig 30a is the sum of six regions. These can be conveniently labelled with the capital letters shown in Fig 30b. We take Left as bLack. Each of the the six regions is an all-small game. Computations described in Chapter 8 at the end of Volume 1 of *Winning Ways* yield the atomic weights shown in Table 3.

The total weight is $-1+_2$. The value of region F is *2, which is remote enough. So the first player can win, even though it is somewhat challenging for Black to do so. More details can be found at www.gustavus.edu/~wolfe/games/clobber.

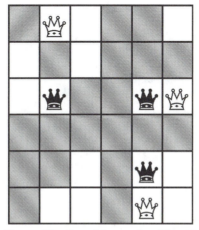

Figure 32. Four Moves after Fig. 29.

Region	A	B	C	D	E	F	Sum
Atomic Weight	$\downarrow*$	$2\Uparrow$	$-2+_2$	$\downarrow*$	-1	0	$-1+_2$

Table 3. Values and Incentives for Fig. 31.

The Konane endgame problem of Fig. 31a virtually decomposes into the sum of ten regions as labelled in Fig 31b. We again take Left as bLack. The values of the regions, ordered by their incentives, are given in Table 4. Both of the possible inter-regional interactions shown in Fig 31b are irrelevant, because in each case, the value of the pair of combined regions is the same as it would be if they were further apart.

Region	A	B	C	D	E	F	G	H	I	J
Value	$\frac{1}{4}\vert*$	$\frac{1}{4}$	$*2$	$\uparrow*$	$-_1$	$*3$	$*$	$-\frac{1}{2}$	$\uparrow*$	$\downarrow$
Δ^R	$\frac{1}{4}*\vert 0$	$-\frac{1}{4}$	$*2,*3$	$\uparrow*$	$-_1$	$*3,*2,*$	$*$	$-\frac{1}{2}$	$\uparrow*$	$\downarrow$
Δ^L	$\frac{1}{4}*\vert 0$	$-\frac{1}{4}$	$*2,*3$	$\downarrow*,\downarrow$	$\{1\vert 0\}+_1$	$*3,*2,*$	$*$	$-\frac{1}{2}$	$\downarrow*,\downarrow$	$\uparrow*$

Table 4. Atomic weights for Fig. 30.

If White plays first, his dominant incentive is on A, after which the value is -1/4 ish, clearly negative and a straightforward win for White. But if Black plays first, her dominant incentive is on E, after which White's dominant response is on E^L. Black's next dominant move is on A, yielding an overall total value which is a positive infinitesimal, so first player can win.

References and Further Reading

Amazons

Elwyn Berlekamp, Sums of $N \times 2$ Amazons, *Inst. Math. Statist. Lect. Notes*, 35(2000) 1–34; MR 2002e:91033.

Michael Buro, Simple Amazons engames and their connections to Hamilton circuits in cubic subgrid graphs, *Proc. 2nd Internat. Conf. Computers and Games (Hamamatsu, 2000), Lecture Notes in Comput. Sci.*, 2063, Springer, Berlin, 2002, 250–261.

T. Hashimoto, Y. Kajihara, H. Iida and J. Yoshimura, An evaluation function for Amazon, *Advances in Computer Games*, 9 (2000)

P. Hensgens and J. Uiterwijk, 8QP wins Amazons tournament, *ICGA J.*, **23** (2000) 179–181.

A. Hollosi, Smart game format for Amazons, 2000, http://www.red-bean.com/sgf/amazons.html

H. Iida and M. Müller, Report on the second open computer-Amazons championship, *ICGA J.*, **23** (2000) 51–54.

Martin Müller and Theodore Tegos, Experiments in computer Amazons, in Richard Nowakowski (ed.) *Games of No Chance*, (Berkeley CA 1994) *Math. Sci. Res. Inst. Publ.*, 29 (1996) Cambridge Univ. Press, Cambridge, UK, 243–260.

R. Rognlie, Play by e-mail server for Amazons 1999, www.gamerz.net/pbmserv/amazonz.html

N. Sasaki and H. Iida, Report on the first open computer-Amazons championship, *ICGA J.*, **22** (1999) 41–44.

Raymond George Snatzke, Exhaustive search in the game Amazons, in Richard Nowakowski (ed.) *More Games of No Chance*, (Berkeley CA 2000) *Math. Sci. Res. Inst. Publ.*, 42 (2002) Cambridge Univ. Press, Cambridge, UK, 261–278.

Checkers

Elwyn Berlekamp, 2002, The 4G4G4G4G4 Problems and Solutions, in Richard Nowakowski (ed.) *More Games of No Chance*, (Berkeley CA 2000) *Math. Sci. Res. Inst. Publ.*, 42 (2002) Cambridge Univ. Press, Cambridge, UK, 231–241.

J. M. Robson, N by N checkers is Exptime complete, *SIAM J. Comput.*, **13** (1984) 252–267.

Jonathan Schaefer, Marion Tinsley: Human perfection at checkers? in Richard Nowakowski (ed.) *Games of No Chance*, (Berkeley CA 1994) *Math. Sci. Res. Inst. Publ.*, 29 (1996) Cambridge Univ. Press, Cambridge, UK, 115–118.

Jonathan Schaefer and Robert Lake, Solving the game of checkers, in Richard Nowakowski (ed.) *Games of No Chance*, (Berkeley CA 1994) *Math. Sci. Res. Inst. Publ.*, 29 (1996) Cambridge Univ. Press, Cambridge, UK, 119–133.

Jonathan Schaefer et al., A World Championship caliber checkers program, *Artificial Intelligence*, 53 (1992) 273–290.

Cherries

M. Albert, S. McCurdy, R. J. Nowakowski and D. Wolfe, Evaluating Cherries (to appear).

Chess

Noam Elkies, On numbers and endgames: combinatorial game theory in chess endgames, in Richard Nowakowski (ed.) *Games of No Chance*, (Berkeley CA 1994) *Math. Sci. Res. Inst. Publ.*, 29 (1996) Cambridge Univ. Press, Cambridge, UK, 135–150.

Aviezri S. Fraenkel and David Lichtenstein, Computing a perfect strategy for $n \times n$ chess requires time exponential in N, *Automata, languages and programming (Akko, 1981)*, Lect. Notes Comput. Sci., 115, 278–293; MR 83c:90182.

Claude E. Shannon, Programming a computer for playing chess, *Philos. Mag.*(7) **41** (1950) 256–275; **11**, 543f.

Lewis Stiller, Multilinear algebra and chess endgames, in Richard Nowakowski (ed.) *Games of No Chance*, (Berkeley CA 1994) *Math. Sci. Res. Inst. Publ.*, 29 (1996) Cambridge Univ. Press, Cambridge, UK, 151–192.

Clobber

E. Demaine, M. Demaine and R. Fleischer, Solitaire Clobber, *J. Theoret. Comput. Sci., Proc. Dagstuhl Seminar Algorithmic Combin. Games* (to appear) prove that a filled rectangular board with at least 2 rows and 2 columns can be reduced to one piece if the number of pieces is not a multiple of 3 and to 2 pieces otherwise. If the board is $1 \times n$ then $n/4$ is the best that can be done.

Michael Albert, J. P. Grossman and David Wolfe (preprint) have results for small positions, including a *2 position. Most of these, and more, appear on Wolfe's website http://www.gustavus.edu/~wolfe /papers or http://www.gustavus.edu/~wolfe/games/clobber

Connect-Four

J. D. Allen, A note on the computer solution of connect-four, in D. N. L. Levy and D. F. Beal (eds) *Heuristic Programming in Artificial Intelligence: the first computer Olympiad*, Ellis Horwood, Chichester, England, 1989.

L. V. Allis, A knowledged-based attack to connect-four. The game is solved. White wins. MSc thesis, Vrije Universiteit Amsterdam, 1988.

Go

David B. Benson, Life in the game of Go, *Information Sci.*, **10** (1976) 17–29; MR 53 #15010.

Elwyn R. Berlekamp and Yonghoan Kim, Where is the "Thousand-dollar Ko"? in Richard Nowakowski (ed.) *Games of No Chance*, (Berkeley CA 1994) *Math. Sci. Res. Inst. Publ.*, 29 (1996) Cambridge Univ. Press, Cambridge, UK, 203–226; MR 97i:90133.

Elwyn R. Berlekamp and David Wolfe, *Mathematical Go: Chilling Gets the Last Point*, A K Peters, Natick, MA, 1994. Also published in paperback, with accompanying software, as *Mathematical Go: Nightmares for the Professional Go Player*, Ishi Press Internat., San Jose CA; Japanese translation available from Toppan Publishers, Tokyo, Japan; MR 95i:90131.

John D. Goodell, *The world of ki*, Riverside Research Press, St. Paul MN, 1958; MR 19 1248n.

Anders Kierulf, Human-computer interaction in the game of Go, in Zbigniew W. Raś and Maria Zemankova. (eds) *Methodologies for intelligent systems, Proc. 2nd Internat. Symp. (Charlotte NC)*, 1987, North-Holland, New York, 1987 481–487; MR90a:68005.

Kim Jin-Bai, On the game of Go, *J. Korean Math. Soc.*, **14** (1977/78) 197–205; **57** #11798; *Kyungpook Math. J.*, **18** (1978) 125–134; **58** #15226; Erratum **19** (1979) 149; MR 81a:90162.

Howard A. Landmam, Eyespace values in Go, in Richard Nowakowski (ed.) *Games of No Chance*, (Berkeley CA 1994) *Math. Sci. Res. Inst. Publ.*, 29 (1996), Cambridge Univ. Press, Cambridge, UK, 227–257; MR 97j:90100.

A. Mateescu, Gh. Păun, G. Rozenberg and A. Salomaa, Parikh prime words and GO-like territories, *J.UCS*, **1** (1995) 790–810 (electronic); *MR* 97b:68113.

Charles Mathews, Teach Yourself GO, Teach Yourself Books, London, 1999.

David Moews, On some combinatorial games connected with Go, PhD thesis, Univ. of California, Berkeley, 1993.

David Moews, Loopy games and Go. in Richard Nowakowski (ed.) *Games of No Chance*, (Berkeley CA 1994) *Math. Sci. Res. Inst. Publ.*, 29 (1996) Cambridge Univ. Press, Cambridge, UK, 259-272; MR 98d:90152.

David Moews, Coin-sliding and Go, *Theoret. Comput. Sci.*, **164** (1996) 253–276; MR 97h:90094.

Martin Müller, Elwyn Berlekamp and Bill Spight, Generalized thermography: algorithms, implementation and application to Go endgames. Technical Report 96-030, ICSI Berkeley, 1996. Also posted at www.cs.ualberta.ca/~mmueller/

Martin Müller and Ralph Gasser, Experiments in computer Go endgames, in Richard Nowakowski (ed.) *Games of No Chance*, (Berkeley CA 1994) *Math. Sci. Res. Inst. Publ.*, 29 (1996) Cambridge Univ. Press, Cambridge, UK, 273–284.

Teigo Nakamura and Elwyn Berlekamp Analysis of Composite Corridors, in Computers and Games. Third International Conference, CG 2002. Editors J. Schaeffer, Y. Bjornsson, M. Müller. To appear in Springer Lecture Notes in Computer Science.

J. M. Robson, The complexity of Go, in R. E. A. Mason (ed) *Proc. Inform. Processing*, 83(1983) Elsevier, Amsterdam, 1983, 413–417.

Shen Ji-Hong, Mathematical modeling problems in the game of Go, (Chinese) *Math. Practice Theory*, **1995** 15–19..

Arthur Smith, *The game of Go, the national game of Japan*, Charles E. Tuttle, Rutland VT and Tokyo, 1956; Originally published 1908, Moffat, Yard, New York, Photographic copy; MR 18 454c.

Stephen Soulé, The implementation of a Go board, *Inform. Sci.*, **16** (1978) 31–40; MR 80d:68117.

William T, Spight, Extended thermography for multiple kos in Go, *Theoret. Comput. Sci.*, **252** (2001) 23–43; MR 2001k:91041.

Takenobo Takizawa, Am application of mathematical game theory to Go endgames some width-two-entrance rooms with and without kos, in Richard Nowakowski (ed.) *More Games of No Chance*, (Berkeley CA 2000) *Math. Sci. Res. Inst. Publ.*, 42 (2002) Cambridge Univ. Press, Cambridge, UK, 107–124.

Edward Thorp and William E. Walden, A computer assisted study of Go on $M \times N$ boards, *Information Sci.* **4** (1972) 1–33; MR 46 #8684; A partial analysis of Go. *Comput. J.*, **7** (1964) 203–207; **33** #2424.

S. Willmott, J. Richardson, A. Bundy and J. Levine, Applying adversarial planning techniques to Go, *Theoret. Comput. Sci.*, **252** (2001) 45–82; MR 2001j:68114.

David Wolfe, Mathematics of Go: chilling corridors, PhD thesis, Univ. of California, Berkeley, 1991.

David Wolfe, Go endgames are PSPACE-hard, in Richard Nowakowski (ed.) *More Games of No Chance*, (Berkeley CA 2000) *Math. Sci. Res. Inst. Publ.*, 42 (2002) Cambridge Univ. Press, Cambridge, UK, 125–136.

Konane

Alice Chang and Alice Tsai, $1 \times n$ Konane: a summary of results, in Richard Nowakowski (ed.) *More Games of No Chance*, (Berkeley CA 2000) *Math. Sci. Res. Inst. Publ.*, 42 (2002) Cambridge Univ. Press, Cambridge, UK, 331–339.

Michael D. Ernst, Playing Konane mathematically: a combinatorial game-theoretic analysis; *UMAP J.*, **16** (1995) 95–121.

Reversi-Othello

M. Buro, Experiments with Multi-ProbCut and a new high-quality evaluation function for Othello, in J. van den Herik and H. Iida (eds.) *Games in AI Research*, Univ. Maastricht, 2000, 77–96.

Gao Xin-Bo, Hiroyuki Iida, Jos W. H. M. Uiterwijk and H. Jaap van den Herik, Strategies anticipating a difference in search depth using opponent-model search, *Computer and games (Tsukuba, 1998)*, *Theoret. Comput. Sci.*, **252** (2001), 83–104; MR 2001j:68111.

Shigeki Iwata and Takumi Kasai, The Othello game on an $n \times n$ board is PSPACE-complete, *Theoret. Comput. Sci.*, **123** (1994) 329–340; MR 95a:68043.

Scrabble

David Wolfe and Susan Hirshberg, All tied up in naughts, in David Wolfe and Tom Rodgers (eds) *Puzzlers' Tribute: A Feast for the Mind*, A K Peters, Natick MA, 2002, pp. 53–58; MR 2003a:00005.

Shogi

Donald F. Beal and Martin C. Smith, Temporal difference learning applied to game playing and the results of application to shogi, *Computer and games* (*Tsukuba*, 1998), *Theoret. Comput. Sci.*, **252** (2001) 105–119; MR 2001j:68106.

Yoshio Hoshi, Kohei Noshita and Keiji Yanai, A new algorithm for solving the cooperative tsume-shogi based on iterative-deepening search (Japanese), *IPSJ J.*, **43** (2002), 11–19.

Ayumu Nagai and Hiroshi Imai, Application of df-pn algorithm to a program to solve Tsume-Shogi problems (Japanese), *IPSJ J.*, **43** (2002) 1769–1777.

Kohei Noshita and Takahito Iida, Generalized solution-sequences in games and enumeration of a certain type of Tsume-Shogi (Japanese) *IPSJ J.*, **43** (2002) 708–713.

E. Ohara, *Japanese chess: the game of Shogi* Charles E. Tuttle, Rutland VT. and Tokyo, 1958. MR 19, 1248o.

Makoto Sakuta and Hiroyuki Iida, AND/OR-tree search for solving problems with uncertainty—a case study using screen-shogi problems (Japanese) *IPSJ J.*, **43** (2002) 1–10.

Masahiro Seo, Hiroyuki Iida and Jos W, H. M. Uiterwijk The PN*-search algorithm: application to tsume-shogi, *Heuristic search in artificial intelligence; Artificial Intelligence*, **129** (2001) 253–277.

Nobusuke Sasaki and Hiooyuki Iida, The study of evolutionary change of Shogi (Japanese). Special issue on game programming. *IPSJ J.*, **43** (2002) 2990–2997.

Hirohisa Seki and Hidenori Itoh, Inducing Shogi heuristics using inductive logic programming, in David Page (ed.) *Inductive logic programming; Proc. 8th Internat. Conf.* (ILP-98) *Madison WI* 1998. Lect. Notes Comput. Sci., 1446 (1998) 155–164, Springer, Berlin, 1998.

Sowing Games

Ethan Akin and Morton Davis, Bulgarian solitaire. *Amer. Math. Monthly*, **92** (1985) 237–250; MR 86m:05014.

L. Victor Allis, Maarten van der Meulen and H. Jaap van den Herik, Proof-number search, *Artificial Intelligence*, **66** (1994) 91–124; MR 95e:68038.

Hans-J. Bentz, Proof of the Bulgarian Solitaire conjectures. *Ars Combin.* **23** (1987) 151–170; MR 88k:05018.

Anders Bjørner and Láslo Lovász, Chip-firing games on directed graphs, *J. Algebraic Combin.*, **1** (1992) 305–328; MR 94c:90132.

Jørgen Brandt, Cycles of partitions, *Proc. Amer. Math. Soc.* **85** (1982) 483–486; MR 83i:05012.

Duane M. Broline and Daniel E. Loeb, The combinatorics of Mancala-type games: Ayo, Tchoukaillon, and $1/\pi$, *UMAP J.*, **16** (1995) 21–36; MR 98a:90166.

C. Cannings and J. Haigh, Montreal solitaire, *J. Combin. Theory Ser. A* **60** (1992) 50–66; MR 93d:90082.

J. Culberson, Sokoban is PSPACE complete, in: *Fun With Algorithms*, Vol. 4 of *Proc. Inform.*, Carleton Scientific, Univ. Waterloo, Ont., 1999, 65–76.

Jeff Erickson, Sowing games, in Richard Nowakowski (ed.) *Games of No Chance*, (Berkeley CA 1994) *Math. Sci. Res. Inst. Publ.*, 29 (1996) Cambridge Univ. Press, Cambridge, UK, 287–297. Used Wolfe's Toolkit (ibid. 93–99) which (Jun 2003) is being replaced by Aaron Siegel's *cgsuite*.

Dwihen Étienne, Tableaux de Young et solitaire bulgare, *J. Combin. Theory Ser. A*, **58** (1991) 181–197; MR 93a:05134.

T. S. Ferguson, Some chip transfer games, *Theoret. Comput. Sci.* (*Math Games*), **191** (1998) 157–171.

Kiyoshi Igusa, Solution of the Bulgarian solitaire conjecture, *Math. Mag.*, **58** (1985) 259–271; MR 87c:00003.

I. Russ, *Mancala games*, Algonac MI, 1984 gives 1200 known variants!

N. J. A. Sloane, My favorite integer sequences, in *Sequences and their applications (Singapore, 1998)*, 103–130; Springer Ser. Discrete Math. Theor. Comput. Sci., Springer, London, 1999; MR 2002h:11021.

Yeh Yeong-Nan, A remarkable endofunction involving compositions, *Stud. Appl. Math.*, **95** (1995) 419–432; MR 97e:05027

Additional Selected References

L. V. Allis, H. J. van den Herik and M. P. H. Huntjens, Go-Moku solved by new search techniques, in *Proc. AAAI Fall Symp. on Games: Planning and Learning*, AAAI Press Tech. Report FS93-02, Menlo Park CA, 1993 1–9.

William N. Anderson, Maximum matching and the game of Slither, *J. Combinatorial Theory Ser. B*, **17**(1974) 234–239.

Vadim V. Anshelevich, The game of Hex: the hierarchical approach, in Richard Nowakowski (ed.) *More Games of No Chance*, (Berkeley CA 2000) *Math. Sci. Res. Inst. Publ.*, 42 (2002) Cambridge Univ. Press, Cambridge, UK, 151–165.

Charles Babbage, *Passages from the Life of a Philosopher*, Longman, Green, Longman, Roberts and Green, London, 1864; reprinted Augustus M. Kelley, New York, 1969, pp. 467–471.

J. P. V. D. Balsdon, *Life and Leisure in Ancient Rome*, McGraw-Hill, New York. 1969, pp. 156 ff.

A. G. Bell, Kalah on Atlas, in D. Michie (ed.) *Machine Intelligence, 3*, Oliver & Boyd, London, 1968, 181–193

A. G. Bell, *Games Playing with Computers*, George Alien & Unwin. London, 1972, pp. 27–33.

Robert Charles Bell, *Board and Table Games from Many Civilizations*, Oxford University Press, London, 1960, 1969.

Richard A. Brualdi, Networks and the Shannon Switching Game, *Delta*, **4**(1974) 1–23.

Cameron Browne, *Hex Strategy: Making the Right Connections*, A K Peters, Ltd. Natick MA, 2000.

Gottfried Bruckner, Verallgemeinerung eines Satzes über arithmetische Progressionen, *Math. Nachr.* **56**(1973) 179–188; M.R. 49#10562.

L. Csmiraz [Csirmaz], On a combinatorial game with an application to Go-Moku, *Discrete Math.* **29**(1980) 19–23.

D. W. Davies, A theory of Chess and Noughts and Crosses, *Sci. News*, **16**(1950) 40–64.

Erik D. Demaine, Martin L. Demaine and David Eppstein, Phutball endgames are hard, in Richard Nowakowski (ed.) *More Games of No Chance*, (Berkeley CA 2000) *Math. Sci. Res. Inst. Publ.*, 42 (2002) Cambridge Univ. Press, 351–360.

H. E. Dudeney, *The Canterbury Puzzles and other Curious Problems*, Thomas Nelson and Sons, London, 1907; Dover, New York, 1958.

H. E. Dudeney, *536 Puzzles and Curious Problems*, ed. Martin Gardner, Chas. Scribner's Sons, New York, 1967.

J. Edmonds, Lehman's Switching Game and a theorem of Tutte and Nash-Williams, *J. Res. Nat. Bur. Standards*, **69B**(1965) 73–77.

P. Erdős and J. L. Selfridge, On a combinatorial game, *J. Combinatorial Theory Ser. B*, **14**(1973) 298–301.

Ronald J. Evans, A winning opening in Reverse Hex, J. Recreational Math. 7 (1974) 189-192.

Ronald J. Evans, Some variants of Hex, *J. Recreational Math.* **8**(1975–76) 120–122.

Edward Falkener, *Games Ancient and Oriental and How to Play Them*, Longmans Green, London, 1892; Dover, New York, 1961.

G. E. Felton and R. H. Macmillan, Noughts and Crosses, *Eureka*, **11**(1949) 5–9.

William Funkenbusch and Edwin Eagle, Hyperspacial Tit-Tat-Toe or Tit-Tat-Toe in four dimensions, *Nat. Math. Mag.* **19** #3 (Dec. 1944) 119–122.

David Gale, The game of Hex and the Brouwer fixed-point theorem, *Amer. Math. Monthly*, **86** (1979) 818–827.

Martin Gardner, *The Scientific American Book of Mathematical Puzzles and Diversions*, Simon & Schuster, New York, 1959.

Martin Gardner, Mathematical Games, *Scientific Amer.*, each issue, but especially **196** #3 (Mar. 1957) 160–166; **209** #4 (Oct. 1963) 124–130; **209** #5 (Nov. 1967) 144–154; **216** #2 (Feb. 1967) 116–120; **225** #2 (Aug. 1971) 102–105; **232** #6 (June 1975) 106–111; **233** #6 (Dec. 1975) 116–119; **240** #4 (Apr. 1979) 18–28.

Martin Gardner, *Sixth Book of Mathematical Games from Scientific American*, W.H. Freeman, San Francisco, 1971; 39–47.

Martin Gardner, *Mathematical Carnival*, W.H. Freeman, San Francisco 1975, Chap. 16.

Ralph Gasser, Solving Nine Men's Morris, in Richard Nowakowski (ed.) *Games of No Chance*, (Berkeley CA 1994) Math. Sci. Res. Inst. Publ., 29 (1996) Cambridge Univ. Press, Cambridge, UK, 101–113.

Solomon W. Golomb and Alfred W. Hales, Hypercube tic-tac-toe, in Richard Nowakowski (ed.) *More Games of No Chance*, (Berkeley CA 2000) Math. Sci. Res. Inst. Publ., 42(2002) Cambridge Univ. Press, 167–182.

J. P. Grossman and Richard J. Nowakowski, One-dimensional Phutball, in Richard Nowakowski (ed.) *More Games of No Chance*, (Berkeley CA 2000) Math. Sci. Res. Inst. Publ., 42 (2002) Cambridge Univ. Press, Cambridge, UK, 361–367.

Richard K. Guy and J. L. Selfridge, Problem S. 10, *Amer. Math. Monthly*, **86**(1979) 306; solution T. G. L. Zetters **87**(1980) 575–576.

A. W. Hales and R. I. Jewett, Regularity and positional games, *Trans. Amer. Math. Soc.* **106** (1963) 222–229; M.R. 26 # 1265.

Heiko Harborth and Markus Seemann, Snaky is an edge-to-edge looser, *Geombinatorics*, **5** (1996) 132–136.

H. Harborth, Snaky is a paving winner, http://bmi1.math.nat.tu-bs.de/preprints/199614.ps MR 97g:05052.

Professor Hoffman (Angelo Lewis), *The Book of Table Games*, Geo. Routledge & Sons, London, 1894, pp. 599–603.

Isidor, Bishop of Seville, *Origines*, Book 18, Chap. 64.

Edward Lasker, *Go and Go-Moku*, Alfred A. Knopf, New York, 1934; 2nd revised edition, Dover, New York, 1960.

Alfred Lehman, A solution of the Shannon switching game, *SIAM J.* **12**(1964) 687–725.

E. Lucas, *Récréations Mathématiques*, Gauthier-Villars, 1882–1894, Blanchard, Paris, 1960.

Carlyle Lustenberger, M.S. thesis, Pennsylvania State University, 1967.

Leo Moser, Solution to problem E773 [1947,281], *Amer. Math. Monthly*, **55**(1948) 99.

Geoffrey Mott-Smith, *Mathematical Puzzles*, Dover, New York, 1954; ch. 13 Board Games.

H. J. R. Murray, *A History of Board Games other than Chess*, Oxford University Press, 1952; Hacker Art Books, New York, 1978; chap. 3, Games of alignment and configuration.

T. H. O'Beirne, New boards for old games. *New Scientist*, **269**(62:01:11).

T. H. O'Beirne, *Puzzles and Paradoxes*, Oxford University Press, 1965.

Hilarie K. Orman, Pentominoes: a first player win, in Richard Nowakowski (ed.) *Games of No Chance*, (Berkeley CA 1994) Math. Sci. Res.Inst. Publ., 29 (1996) Cambridge Univ. Press, 339–344.

Ovid, *Ars Amatoria*, ii, 208, iii, 358.

Oren Patashnik, Qubic: $4 \times 4 \times 4$ tic-tac-toe, *Math. Mag.*, **53** (1980) 202–216; MR 83e:90169.

Jerome L. Paul, The q-regularity of lattice point paths in R^n, *Bull. Amer. Math. Soc.* **81**(1975) 492; Addendum, ibid. 1136.

Jerome L. Paul, Tic-Tac-Toe in n dimensions, *Math. Mag.* **51**(1978) 45–49.

Jerome L. Paul, Partitioning the lattice points in R^n, *J. Combin. Theory Ser. A*, **26**(1979) 238–248.

Harry D. Ruderman, The games of Tick-Tack-Toe, *Math. Teacher* **44** (1951) 344–346.

Sidney Sackson, *A Gamut of Games*, Random House, New York, 1969.

John Scarne, *Scarne's Encyclopedia of Games*, Harper and Row, New York, 1973.

Fred. Schuh, *The Master Book of Mathematical Recreations*, trans. F. Göbel, ed. T. H. O'Beirne, Dover, New York, 1968; ch. 3, The game of Noughts and Crosses.

R. Uehara and S. Iwata, Generalized Hi-Q is NP-complete, *Trans. IEICE*, **E73** (1990) 270–273.

Index to Volumes 1–3

Grossman, J. P., 755, 757, 763, 768
ground (=earth), 193, 550, 551
grounding cycles, 193, 196
grown-up picture, 43
Grundy
    scale 87, 90, 91, 94, 96, 101, 398–399, 406
    Skayles, 91
Grundy, Mrs., 310
Grundy, Patrick Michael, 42, 56, 79, 117, 220,
    221, 333, 417, 444, 454
Grundy's Game, 15, 96, 112, 434, 439–440,
    444, 690
    misère, 416–420
    wild animals, 431
Grunt, 472, 473, 481
Guiles = ·15, 94, 101, 103, 436, 444
Guy, Michael John Thirian, 464, 520, 535, 621
Guy, Richard Kenneth, 18, 89, 99, 109, 117,
    118, 144, 489, 539, 621, 640, 768

hack, 675, 684, 693, 695
Hackenbush
    Blue-Red, 1–6, 17, 20, 23, 27, 28, 77, 78,
        197, 211–217
    Childish, 43, 52, 157, 237
    Double, 343
    Green, 33, 34, 189–196
    Hotchpotch, 29, 33, 36, 37, 47, 66, 67, 197–
        210, 218–222, 237, 242–246, 251
    infinite, 327–332, 344, 345
    is hard!, 211, 217
    loopy, 343–345
    number system, 78
    picture, 1–5, 189–195
    string, 22, 23, 77, 78, 194, 195, 327–331
    von Neumann, 606
Hackenbush companion, 693
Hackenbush, flowers, 675, 695
hacking toll, 686
Haigh, J., 766
Hala-Tafl, 666
Hales, A. W., xvii, 740, 741, 768
half-move, 4, 7, 9, 19, 20
half-off, 356
half-on, 356
half-tame, 414, 423, 435, 444
halving, 620
    of nim-values, 195

handouts, 547
Hanner, Olof, 188
Harary, Frank, 748
Harborth, H., 768
hard game, 211, 217, 222, 223
hard problems, 223–225
hard redwood bed, 217, 222
hard-headed, 179
hardness, 211, 217, 223, 225
Hare and Hounds, 711–729
harmless mutation, 562, 564
Harrocks, 517
Harry Kearey, 318
Hashimoto, T., 762
Hawaii, 16
Hawaiian checkers, 759
head
    animals' 205
    girl's 193, 196
    losing your, 222
    severed, 205, 220–222
    shrunken, 220–221
heaps, see also Nim-heaps
heap, see also Nim-heaps
    black, 532
    colored, 532
    grey, 532
    quiiddity, 534
    white, 532
hearts, 359
heat, 125, 132, 145, 299
heat, latent, 307
heating, 167, 173, 690
height, 675, 686, 687, 698, 703
Hein, Piet, 744
Hensgens, P., 762
hereditarily tame, 425
heuristic discussion, 158, 159
Hex, 226, 744, 767, 768
hexadecimal games, 116, 117
hexagon, 744
Hexomino, 748
Hi, 609, 610, 617, 631
hi, 355, 356
Hickerson, Dean, 51
Hickory, Dickory, Dock, 521
high atomic weights, 706